D1224229

ALLE·ZEIT·WACH
1842

Assem Deif

Sensitivity Analysis in Linear Systems

With 15 Figures

Springer-Verlag Berlin Heidelberg NewYork
London Paris Tokyo

Prof. Assem Deif
Department of Engineering Mathematics
Cairo University
Giza/Egypt

ISBN 3-540-16312-3 Springer-Verlag Berlin Heidelberg New York
ISBN 0-387-16312-3 Springer-Verlag New York Heidelberg Berlin

Library of Congress Cataloging in Publication Data:
Deif, Assem:
Sensitivity analysis in linear systems / Assem Deif.
Berlin, Heidelberg, New York: Springer, 1986.
ISBN 3-540-16312-3 (Berlin . . .)
ISBN 0-387-16312-3 (New York . . .)

Printing: Sala-Druck, Berlin
Bookbinding: Schöneberger Buchbinderei, Berlin
2161/3020-543210

To Shawki Deif
with
admiration

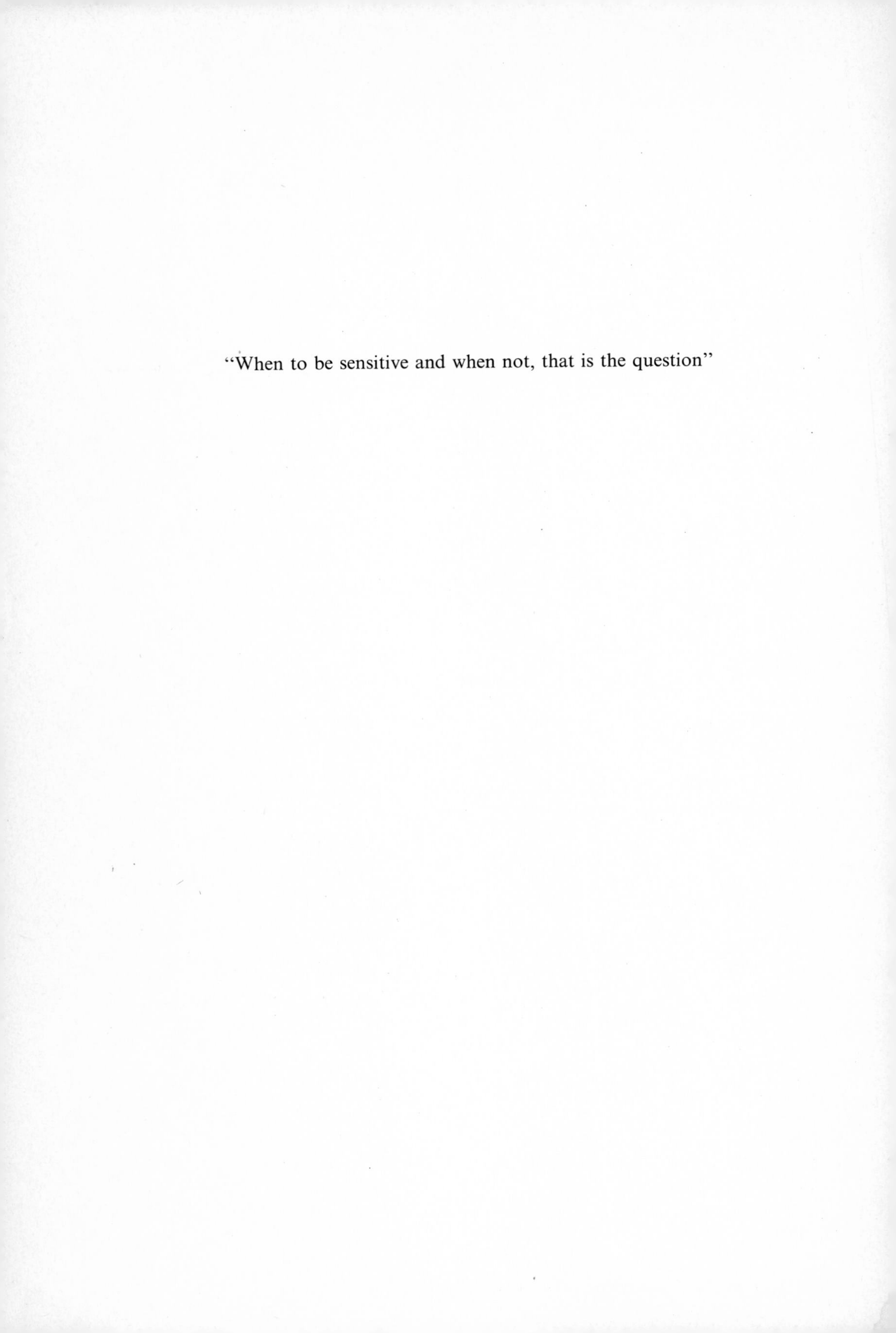

"When to be sensitive and when not, that is the question"

Preface

A text surveying perturbation techniques and sensitivity analysis of linear systems is an ambitious undertaking, considering the lack of basic comprehensive texts on the subject. A wide-ranging and global coverage of the topic is as yet missing, despite the existence of numerous monographs dealing with specific topics but generally of use to only a narrow category of people. In fact, most works approach this subject from the numerical analysis point of view. Indeed, researchers in this field have been most concerned with this topic, although engineers and scholars in all fields may find it equally interesting.

One can state, without great exaggeration, that a great deal of engineering work is devoted to testing systems' sensitivity to changes in design parameters. As a rule, high-sensitivity elements are those which should be designed with utmost care. On the other hand, as the mathematical modelling serving for the design process is usually idealized and often inaccurately formulated, some unforeseen alterations may cause the system to behave in a slightly different manner. Sensitivity analysis can help the engineer innovate ways to minimize such system discrepancy, since it starts from the assumption of such a discrepancy between the ideal and the actual system.

All in all, methods of mathematical optimization rely one way or the other on relative sensitivities, under a different title in each, ranging from gradient methods to model tracking or self-learning systems. Even the simple task of fitting data to a curve usually involves sensitivity calculations. As for social scientists, economists, as well as for many other disciplines, sensitivity and perturbation techniques can provide valuable information about the amount of inaccuracy in the behaviour of a model as related to the inaccuracies in the system's data. If the data gathered by field study or experimental testing falls within certain tolerance limits, the tolerances may well be amplified and widened in the output results obtained. The question might then arise as to how uncertain the results are — or how unrealiable — in relation to the data's uncertainties. In this instance, perturbation analysis can provide valuable information about regions of compatibility and admissibility of solutions. An alternate use might also be to determine the allowable data tolerances in a parameter for the results to sustain a certain level of accuracy; and so forth.

As rewarding a subject as it may be, sensitivity analysis still imposes a tedious job when it comes to organizing a text on it. As the text is intended to serve a wide audience, applications of various kinds had to be included, and a huge effort had to be devoted to ensuring as comprehensive a discussion as possible of the area of linear systems. Some texts have tried to attract the widest readership by choosing some topics of the linear systems and some of the nonlinear ones. Practical experience has shown that such texts have little to add to the real user's knowledge, and only serve to provide an idea about the subject. The present line of approach is stronger, in that it provides not only practical applications but mainly a coherent mathematical justification that brings still further applications within range.

Perturbation techniques and sensitivity analysis are of course no new terms, nor are they recently explored fields. As a mathematical discipline, however, a unified body of knowledge rather than an elementary application, they are rather young. Only during this last century have celestial mechanics witnessed an era of rapid progress; the three-body problem — in contrast with Newton's two body problem — becoming the new challenge. Workers in the field opted to consider this third body as a perturbation in the field. In this context also, Lagrange's ingenious method of variation of elements was introduced, and Poincaré's theory of asymptotic expansions enabled the summing up of a few terms of a divergent series to yield almost exactly its sum.

Perturbation theory in linear algebra is an even more recent branch. In 1948, Turing's famous paper triggered interest in the problem of sensitivity of solutions of linear equations to round-off errors. In this paper, Turing laid down the definition of a condition number by which a small input error in the data can be drastically amplified in the solution. Numerical analysts then acknowledged this number as the major factor affecting computational accuracy, and have tried since then to control it while working out any new numerical procedure. But an ill-conditioned system can only be cured up to a certain extent, and no matter how cunning, skilled or elaborate one is, Turing's number or a variant of it will eventually hinder our illusion.

This work covers the subject of sensitivity analysis as related to linear equations. At first, the plan was to furnish one for linear systems in general, but as the work grew it was found impossible to survey the whole theory in one volume. It was then decided to release the available material as it constitutes a unified body of knowledge. Naturally, we started with the first basic problem in linear systems, that of linear equations, for we shall need many of the results if we are to proceed further. As the reader might have noticed, it was not in our plan to either survey or compare the different numerical methods for solving the equations. For this, he may refer to current literature in numerical analysis, which is plentiful. Rather, our task was only concerned with the problem of executing perturbation analysis of the equations, while making reference to some relevent applications.

The desire on the part of the author to provide such a text grew up incidentally from his training in engineering and related disciplines. Workers in theses fields sometimes encounter problems in which it is important to perceive the accuracy of the results when the input data is subject to uncertainty or errors. This they seek to determine irrespective of any numerical treatment of the problem. They may also wish to pursue a sensitivity analysis of their models under newly varying conditions. It was this philosophy that inspired the writing of the text and which led the author to prepare the rest of the manuscript. This explains why rounding errors are incorporated into the larger context of perturbations, and why no effort has been made to discuss error analysis from the view point of comparing different numerical strategies. In any case, there is a vast literature on this subject alone. And instead of making entangling the reader so overwhelmingly in the different numerical procedures for solving the equations, it concentrates on deepening his working knowledge in this fruitful area. This does not at all mean that numerical analysts cannot profit from it. On the contrary, many up-to-date error bounds have been included and compared. Furthermore, criteria for validating the solutions and ways of improving them are to be found therein.

The text is therefore a survey and a working knowledge book on perturbation techniques and sensitivity analysis as applied to linear equations and linear programming. It uses a moderate language which appeals to engineers and applied mathematicians. Many workers in various disciplines will find it equally valuable. And as a text, it can easily fit — from experience — into a first course on the subject to be taught in one lecture per week over one semester. As to its rigor, it will soon be realized that there is some overlap in levels, in the sense that some knowledge is standard while some is culled from original papers. Our intention was to encourage readers of different backgrounds and training to approach the subject.

The book consists of five chapters, each related to a specific case so as to make it self-contained. Hence, it will be found that it is not strictly necessary to read the text from the beginning. In other words, it looks as though each chapter is independently written for the reader interested in one specific subject. And we certainly make no claim to completeness in any one of them.

In conclusion, the author feels that such a subject deserves a global coverage which communicates efficiently with the different audiences. And with this specific idea in mind, we hope to have fulfilled this aim and to have filled a gap in the available literature.

Indeed, a word of gratitude should be addressed to all who have contributed to the appearance of this book. Prof. P. Spellucci at the Technische Hochschule in Darmstadt, Prof. T. Yamamoto at Ehime University and Dr. J. Garloff at the Universität Freiburg read various parts of the manuscript and made several suggestions which greatly improved the text and rescued me from many blunders. Any remaining pit-falls become naturally those of the author. Dr. J. Rohn at Charles University provided the author with a mathematical notion which helped in some proofs in section 2.6. The author is also grateful to Prof. N. Makary and Prof. E. Rasmy for fruitful discussions related respectively to sections 4.5 and 5.3, and also to Dr. A. Hussean and Dr. T. El-Mistikawy who helped in calculations of sections 3.4 and 4.3; all are based at Cairo University. The whole project of writing this text would have indeed been foundered without permission from Cairo University to allow me to visit the University of California-Berkeley on my sabbatical leave, thus making use of its rich library. The same also applied to the Alexander von Humboldt Foundation, which gave me the opportunity to visit the Technische Hochschule in Darmstadt, enabling me also to profit from the staff there as well as its different libraries. I would like also to take the opportunity of thanking Prof. Å. Björck for his valuable comments on part of the manuscript. My sincere thanks are also addressed to Eng. A. Assaad for editing the text; the effort he put into this work is greatly appreciated. Finally, I am grateful to my wife for allowing me ample time on the manuscript.

Cairo, August, 1986 A. S. Deif

Table of Contents

Chapter 1

Perturbation of Linear Equations

1.1 Introduction

This chapter will discuss the behaviour of the system of linear simultaneous equations

$$Ax = b$$

when the matrix A and the vector b are subjected to small order perturbations ΔA and Δb respectively. The problem then becomes

$$(A + \Delta A)(x + \Delta x) = b + \Delta b$$

and we are mainly concerned with studying the deviation Δx of the solution with the perturbation. Such an exercise is called *sensitivity analysis*, for the extent of the deviation Δx relative to ΔA and Δb defines the *sensitivity* of the system. A highly sensitive system is roughly one where a relatively large deviation Δx is incurred by small perturbations ΔA and Δb. As we shall see, highly sensitive systems are generally to be avoided in practice. They are referred to as *ill-conditioned*, for a highly sensitive system would yield large variations in the results for only small uncertainties in the data. To clarify this fact, let us study the behaviour of the system of equations

$$x + y = 2$$
$$0.49x + 0.51y = 1$$

— representing a pair of straight lines intersecting at $x = 1$, $y = 1$ — when a small term ε is added to the equations. Surprisingly enough, the set of equations obtained, namely

$$x + y = 2 + \varepsilon$$
$$0.49x + 0.51y = 1$$

represents a pair of straight lines meeting at a point (x, y) given by $x = 1 + 25.5\varepsilon$; $y = 1 - 24.5\varepsilon$, and rather distant from $x = 1$, $y = 1$. Here, a small change of order ε in the equations has produced a change of 25ε in the solution. This system has definitely a high sensitivity to perturbations in its matrix A and vector b.

It is indeed worth noting that sensitivity analysis is usually performed after a problem has been solved for x. It is not intended to find an adjusted solution for the system

$$(A + \Delta A)\, x = b + \Delta b$$

for then, the effect of the perturbations ΔA and Δb on x will have remained unexamined. Rather, it aims at representing Δx as a function of the perturbations ΔA and Δb, in order to elicit their effect on the original solution. For the forementioned example, this function would be given by

$$\frac{\Delta x}{x} \cong 50 \frac{\Delta b_1}{b_1}$$

At this stage, let us examine the origins of the perturbations ΔA and Δb of a system of linear equations and their possible practical meanings. In the field of economy, social sciences, etc . . ., system perturbations usually stem from the lack of precision of the data collected through field observations, this lack being termed *uncertainty of the data*. For the engineer they may be intentionally induced into one of the design parameters to investigate the behaviour in service of the designed system, be it an electrical system, a chemical plant or a building's structure. For the mathematician, perturbations might appear as the result of truncating some infinite series, say π or e.

Finally, coming last though not least, perturbations represent in numerical analysis the effect of round-off errors. Such errors might arise during data reading, or in the course of computation, at the level of an intermediate result. Of all disciplines, numerical analysts have been most interested in this area for the sake of precision in the results they obtain.

This text will not assign to ΔA and Δb any one of the foregoing interpretations, dealing with them in a most general manner through their symbols. This approach does not void the symbols from their factual content. Instead, it keeps them in a form so general that they can account at the same time for all perturbations inherent to — or induced in the system. On some occasions, some interpretation or the other will be emphasized, only for the sake of illustrating the concept. Being far from specialization, this text is not intended for the sole use of numerical analysts. For the more specialized application of error analysis to various computational algorithms, the reader is referred to Wilkinson's treatises (1963, 65), which form only a part of the literature on the subject.

However, it may be of interest to illustrate here how rounding-off can be accounted for as a perturbation. Let us therefore consider the solution for $x = (x_1, x_2)^T$ of the system

$$\begin{pmatrix} 2 & 1 \\ 1 & 2 \end{pmatrix} \begin{pmatrix} x_1 \\ x_2 \end{pmatrix} = \begin{pmatrix} 2 \\ -1 \end{pmatrix}$$

when performed on a four-digit machine with fixed-point arithmetic. The solution comes as

$$\hat{x}_1 = 1.667\,, \qquad \hat{x}_2 = -1.333$$

This is only an approximation to the exact solution, since substitution into the set of equations yields the residual

$$r = A\hat{x} - b = (0.001, 0.001)^T$$

The error residual is due to the machine's approximation of the solution which is more exactly

$$x_1 = 5/3 \qquad x_2 = -4/3$$

Alternatively, one can state that the vector $\hat{x}$ obtained is an exact solution of the perturbed system

$$A\hat{x} = b + \Delta b$$

where $\Delta b = r$.

In any case, whether ΔA and Δb account for errors due to rounding, truncation or inaccuracy in the input data, the solution obtained will still deviate from the ideal case. For the above example, the solution's deviation can be expressed by the vector

$$\begin{aligned}\Delta x &= (1.667 - 5/3\,, \qquad -1.333 + 4/3)^T \\ &= (0.001/3\,, \qquad 0.001/3)^T\end{aligned}$$

Therefore, the value 0.001/3 could be taken as a measure of the deviation, or, alternatively, the sum of deviations in both variables, namely 0.002/3, could be chosen, etc ... In general, how to measure this deviation and in what form is one question, the answer to which necessitates the introduction of the concept of norm.

1.2 Norms of Vectors and Matrices

The norm of a vector is introduced here to provide a measure of the vector's magnitude exactly analogous in concept to that of absolute value for a complex number. The *norm* of a vector x, denoted $\|x\|$, is a non-negative scalar function of x satisfying the following set of axioms

$$\begin{aligned}&\|x\| > 0\,, \qquad \forall x \neq 0 && \text{(Positivity)} \\ &\|cx\| = |c| \cdot \|x\| && \text{(Homogeneity)} \\ &\|x + y\| \leqq \|x\| + \|y\| && \text{(Triangular inequality)}\end{aligned}$$

In general, a scalar function $\|\cdot\|$ satisfying the above axioms qualifies as a norm. The reader can exercise in showing that the following function, termed the *Hölder norm*, and stated as

$$\|x\|_p = \left(\sum_{i=1}^{n} |x_i|^p\right)^{1/p}, \qquad p \geqq 1$$

qualifies as a norm of x. For a proof of the validity of the triangular inequality for this norm, the reader is referred to Deif (1982) for p assuming integer values. For p assuming general values, the reader may refer to Beckenback and Bellman (1965). In the special case where $p = 2$, the general norm function yields the well known Euclidean norm $\|\cdot\|_2$ or $\|\cdot\|_E$, which gives the length of a vector in analytical geometry. Other widely used forms of the norm function are

$$\|x\|_1 = \sum_i |x_i|, \qquad \text{for} \quad p = 1$$

and

$$\|x\|_\infty = \max_i |x_i|, \quad \text{for} \quad p \to \infty$$

Applying each of the foregoing versions of norm to a vector given by

$$x = (1, -2, 3 + i)^T, \qquad i = \sqrt{-1}$$

would yield

$$\begin{aligned} \|x\|_1 &= 1 + 2 + \sqrt{10} = 3 + \sqrt{10} \\ \|x\|_E &= \sqrt{1 + 4 + 10} = \sqrt{15} \\ \|x\|_\infty &= \sqrt{10} \end{aligned}$$

Likewise, the norm of a square matrix A is defined as a nonnegative scalar function noted $\|A\|$ and satisfying the following axioms

$$\begin{aligned} &\|A\| > 0, \qquad \forall A \neq 0 \\ &\|cA\| = |c| \, \|A\| \\ &\|A + B\| \leqq \|A\| + \|B\| \\ &\|AB\| \leqq \|A\| \, \|B\| \end{aligned}$$

Again, many functions could be found that qualify as matrix norms, according to the above rules, e.g.

$$\|A\|_F = \sqrt{\sum_{i,j} |a_{ij}|^2} \qquad \text{(Frobenius norm)}$$

$$\|A\|_M = n \cdot \max_{i,j} |a_{ij}| \qquad \text{(Maximum norm)}$$

$$\|A\|_1 = \max_j \sum_i |a_{ij}| \qquad \text{(Column norm)}$$

$$\|A\|_\infty = \max_i \sum_j |a_{ij}| \qquad \text{(Row norm)}$$

Some have even more properties than those described by the axioms. The last two norms defined above — Row and Column norms — may for instance be *subordinated* to a corresponding vector norm; that is for every matrix A one can always find a vector x such that

$$\|Ax\| = \|A\| \, \|x\|$$

(cf. Young and Gregory (1973); Deif (1982)). Many matrix norms are *consistent* with vector norms, e.g. the Frobenius matrix norm with the Euclidean vector norm, meaning that they satisfy the relation

$$\|Ax\| \leqq \|A\| \, \|x\|$$

However, the equality part of a consistency relation is only verified for specific configurations of A and x, whereas for the Row and Column matrix norms, each matrix A can be subordinated to at least one vector norm $\|x\|$.

In virtue of this additional property, subordination to a vector norm, the row and column norms — and all norms analogous in this respect — are termed *bounds*, or better still, *least upper bounds* (lub). This means that

$$\|A\|_p = \text{lub}_p(A) = \sup_x \frac{\|Ax\|_p}{\|x\|_p}$$

In the cases where $p = 1, \infty$ in the above relation, we obtain respectively the formentioned Column and Row norms. As an example, if

$$A = \begin{bmatrix} 1 & 0 & -1 \\ -3 & 5 & 4 \\ 2 & -2 & 0 \end{bmatrix}$$

then

$$\|A\|_1 = 7, \qquad \|A\|_\infty = 12$$

The derivation of lub(A) necessitates the use of a vector x, whence some authors refer to it as the *induced matrix norm*. If we consider the Frobenius norm, noted $\|A\|_F$ (and held as the matrix analogue of the Euclidean length of a vector), as compared to its corresponding induced norm-version (called the *spectral norm*) given by

$$\|A\|_2 = \sqrt{\lambda_{\max}(A^*A)}$$

where $\lambda_{\max}$ stands for the largest eigenvalue, we find that

$$\|A\|_2 \leqq \|A\|_F$$

This characteristic makes induced matrix norms widely used in relation to error analysis, as they set tighter bounds.

Usually, authors do not differentiate in notation between both types of norms, unless the need rises for the exclusive use of one. In this text, the analytical expressions derived will be valid for nearly all norms, the bounds set by their numerical values being tighter or looser according to the type of norm used. Furthermore, as norms were devised to quantify and compare magnitudes of vectors and matrices, we may settle as analysis proceeds for using the popular lazy quotation: for some norm — as many authors incidentally do. Also, $\|A\|$ will denote any matrix norm, including the cumbersome notation lub (A), further replaced by $\|A\|_1, \|A\|_2, \dots, \|A\|_\infty$.

These definitions of norms, already so well known and frequently used in functional analysis, were employed in 1950 by Faddeeva, in the context of proofs of convergence. Faddeeva defined vector and matrix norms independently, linking them with the concepts of consistency and subordination. The Frobenius norm is called the *absolute value of the matrix* by Wedderburn (1934), who in turn traces the idea back to Peano.

Another type of norm, the Dual norm, was introduced by von Neumann (1937) to be treated axiomatically by many authors, see for instance Stoer (1964). The *Dual norm* of a vector u, noted $\|u\|^D$, is defined by

$$\|u\|_p^D = \sup_x \frac{|\langle u \cdot x \rangle|}{\|x\|_p}$$

For example, for the vector

$$u^T = (1, -2, 3 + i), \qquad i = \sqrt{-1}$$

the dual norm is given by

$$\|u\|_1^D = \sup_x \frac{|\langle u \cdot x \rangle|}{\sum_i |x_i|} = \sup_x \frac{|x_1 - 2x_2 + (3 + i)x_3|}{|x_1| + |x_2| + |x_3|}$$

$$= \sqrt{10}, \qquad \text{taking } x_1 = 0; \qquad x_2 = 0; \qquad x_3 = 1$$

On the other hand

$$\|u\|_\infty^D = \sup_x \frac{|\langle u \cdot x \rangle|}{\max_i |x_i|} = 1 + 2 + \sqrt{10} = 3 + \sqrt{10}$$

taking $x_1 = 1$; $x_2 = -1$; $x_3 = (3 - i)/\sqrt{10}$; $i = \sqrt{-1}$.

This concept of Dual norms is no more than a direct application of Hölder's inequality, stated as

$$\sum_{i=1}^n |u_i x_i| \leq \left(\sum_{i=1}^n |u_i|^p\right)^{1/p} \cdot \left(\sum_{i=1}^n |x_i|^q\right)^{1/q}$$

where u_i and x_i are two sets of numbers; $i = 1, 2, \dots, n$; and

$$1/p + 1/q = 1, \qquad p \geqq 1$$

In fact, if we allow x_i to vary, there will surely exist a homogeneous configuration of x for which the equality strictly holds. In that case $\|u\|_1^D = \|u\|_\infty$ and vice-versa.

What is most interesting for our purposes is however the application of the concept of Dual norm to find the value of lub (uv^*), uv^* being a matrix

$$\operatorname{lub}(uv^*) = \sup_x \frac{\|uv^*x\|}{\|x\|} = \sup_x \frac{|v^*x|\ \|u\|}{\|x\|} = \|u\|\ \|v\|^D$$

Other similar dual matrix norms can also be derived. The reader interested in further information is referred to Stoer (1964), who also deduced some further properties of dual norms as compared to usual norms.

Returning to the system of linear equations

$$Ax = b$$

we recall having stated that solving for x with an accuracy depending on the used computing machine always yields the residual vector r

$$r = A\hat{x} - b$$

This vector r should in some way give an indication of the accuracy of $\hat{x}$. Writing

$$A(x + \Delta x) - b = r$$

$x = A^{-1}b$ being the system's exact solution, we get

$$A\,\Delta x = r$$

or

$$\Delta x = A^{-1}r$$

i.e.

$$\|\Delta x\| = \|A^{-1}r\| \leqq \|A^{-1}\|\ \|r\|$$

so that

$$\frac{\|\Delta x\|}{\|x\|} \leq \frac{\|A^{-1}\|\ \|r\|}{\|A^{-1}b\|} \leq \frac{\|A^{-1}\|\ \|r\|}{\|A\|^{-1}\ \|b\|} = \|A\|\ \|A^{-1}\| \frac{\|r\|}{\|b\|}$$

From this, we can conclude that the magnitude of the relative error in the solution x is bounded by the norm of the residual vector r times the quantity $\|A\|\ \|A^{-1}\|$. This latter quantity is termed the *condition number* of A (noted cond (A)). It plays a vital role in assessing the numerical stability of algorithms, and deserves to be discussed separately. For the time being, let us just clarify the concept by a practical example. Considering the same example discussed before, we will solve

$$\begin{bmatrix} 2 & 1 \\ 1 & 2 \end{bmatrix} \begin{bmatrix} x_1 \\ x_2 \end{bmatrix} = \begin{bmatrix} 2 \\ -1 \end{bmatrix}$$

using a four-digit machine with fixed-point arithmetic. Then

$$\hat{x}_1 = 1.667\,, \qquad \hat{x}_2 = -1.333$$

and

$$r = A\hat{x} - b = \begin{bmatrix} 2 & 1 \\ 1 & 2 \end{bmatrix} \begin{bmatrix} 1.667 \\ -1.333 \end{bmatrix} - \begin{bmatrix} 2 \\ -1 \end{bmatrix} = \begin{bmatrix} 0.001 \\ 0.001 \end{bmatrix}$$

Meanwhile

$$A^{-1} = \begin{bmatrix} 2/3 & -1/3 \\ -1/3 & 2/3 \end{bmatrix}$$

giving

$$\text{cond}\,(A) = \|A\|\,\|A^{-1}\| = 3$$

Hence, using the l_1-norm, we get

$$\frac{\|\Delta x\|}{\|x\|} \leqq \|A\|\,\|A^{-1}\|\frac{\|r\|}{\|b\|} = 3\times 0.002/3 = 0.002$$

Now if we use the exact value of x

$$\Delta x = (0.00033\,\ldots\,,\ 0.00033\,\ldots)^T$$

we get

$$\frac{\|\Delta x\|}{\|x\|} = 0.00033\ldots\times 2/3 = 0.00022\ldots$$

which is of course smaller than the foregoing upper bound. In both cases, the error was small, because — as we will be explaining shortly — *A is well-conditioned.*

Furthermore, we notice that the error in both $\hat{x}_1$ and $\hat{x}_2$ does not exceed 5×10^{-4}. This is simply due to the fact that we used a three-decimal-place precision with the third decimal place rounded according to whether the fourth decimal place is greater or smaller than 5. In this specific example, the exact solution is

$$x_1 = 5/3 = 1.66666\ldots$$
$$x_2 = -4/3 = -1.333333\ldots$$

which yielded after rounding-off

$$\hat{x}_1 = 1.667 \qquad \hat{x}_2 = -1.333$$

By cancelling the fourth decimal place without knowing its value, we have induced an error of 0.0005 at most. For a number displayed with t decimal places, the error would likewise not exceed $5 \times 10^{-t-1} = \frac{1}{2} \times 10^{-t}$. Most calculating machines use floating-point arithmetic for minimal error and better accuracy. The reason behind this becomes clear when a number like 125.7235124 is to be represented in a ten-digit machine. The inherent error does not exceed 5×10^{-8}, but its absolute value is variable, depending on the place of the decimal point. For numbers between 1 and 9.99 ... inclusive, the error will not exceed 5×10^{-10}; however it is not so readily determinable for other numbers. The introduction of floating-point arithmetic simplified this issue by setting the accuracy of the machine itself. Any number is stored as

$$c \times 10^b, \qquad 10 > |c| \geqq 1 \qquad \text{(normalized floating-point: } 1 > |c| \geqq .1)$$

The number c is called the *mantissa*, and has as many digits as the machine itself. The number b is the *exponent*. The figure set forth before can thus be represented as 125.7235124 — the error being 5×10^{-8} — or as 1.257235124×10^2 — the error being now $5 \times 10^{-10} \times 10^2 = 5 \times 10^{-8}$. The accuracy of the machine is to the nearest 5×10^{-10}, and for a number a, the error is at most as large as $5 \times 10^{-10} \times |a|$. For a machine with a t-digit mantissa, the error becomes $5 \times 10^{-t} \times |a|$.

Now, supposing a matrix A is to be be processed on the computer, what would be the maximum error in the norm of A due to rounding-off? In other words, what is the bound for $\|A + \Delta A\| - \|A\|$ due to rounding-off? Since the error in a_{ij} is less than $5 \times 10^{-t} \times |a_{ij}|$ (note that sometimes only part of the machine's mantissa is displayed), then $\|\Delta A\| \leqq 5 \times 10^{-t} \|A\|$. While assuming such error to be less than $5 \times 10^{-t} \times \max_{i,j} |a_{ij}|$ for $a_{ij} \neq 0$ and zero for $a_{ij} = 0$ (see Rice (1981), p. 136), we get

$$|\|A + \Delta A\| - \|A\|| \leq \|\Delta A\| \leq 5 \times 10^{-t} \times n \times \max_{i,j} |a_{ij}|$$

But we have that

$$\|A\| \geq \max_{i,j} |a_{ij}|$$

whence

$$\left| \frac{\|A + \Delta A\| - \|A\|}{\|A\|} \right| \leq 5 \times 10^{-t} \times n$$

For instance, a matrix A of order n processed on an HP-15C calculator — having a ten-digit mantissa — has an error $\|\Delta A\|$ less than $5 \times 10^{-10} \times n\|A\|$ (i.e. less than $10^{-9} n\|A\|$).

1.3 Condition Number and Nearness to Singularity

The condition number of square matrix A, noted cond (A), is a nonnegative real scalar given by

$$\text{cond}(A) = \|A\| \|A^{-1}\|$$

When $\|A\|$ is chosen as the lub (A), it is easily shown that

$$\text{cond}\,(A) = \sup_{u,v} \frac{\|Au\|}{\|Av\|}$$

with $\|u\| = \|v\| = 1$

Geometrically, this equals the ratio between the longest and shortest distances from the origin to points located on the surface

$$y = Ax$$

and such that $\|x\| = 1$. Hence, cond (A) is always greater than unity.

To visualize the relationship between the condition number and the error in the solution of a perturbed system of linear equations, let us again consider the exaggerated example in Sect. 1.1, namely

$$x + y = 2$$

$$0.49x + 0.51y = 1$$

The equations represent a pair of straight lines intersecting at $x = 1$, $y = 1$. Now for some perturbation ε in the right-hand side of the first equation, the equations become

$$x + y = 2 + \varepsilon$$

$$0.49x + 0.51y = 1$$

the solution for which comes as

$$x = 1 + 25.5\varepsilon$$
$$y = 1 - 24.5\varepsilon$$

The original system is termed ill-conditioned (in contrast with well-conditioned) in the sense that a small change of order ε in the equations has induced a change of 25 ε in the solution. This is due to the acuteness of the angle between the two lines. For this example,

$$A = \begin{bmatrix} 1 & 1 \\ 0.49 & 0.51 \end{bmatrix}, \qquad A^{-1} = \begin{bmatrix} 25.5 & -50 \\ -24.5 & 50 \end{bmatrix}$$

yielding cond $(A) \cong 150$. This large condition number is associated with the relatively large error in the result. Usually, the larger the condition number, the smaller the determinant is. As det $(A) \to 0$, we have cond $(A) \to \infty$. A matrix is therefore called ill-conditioned if it is near to singular, as far as arithmetic precision is concerned. In the output of a computer, we might thence encounter the message *matrix is singular or ill-conditioned*, since the machine might not be able to tell the difference unless it uses a very-high-precision arithmetic — especially with small valued determinants or very poor pivots. In both cases, whether the matrix is very ill-conditioned or simply singular, the results obtained are far from being realistic.

To investigate how the condition number can give a measure of nearness to singularity, we will borrow the following example from the HP-15C advanced matrix functions library. If

$$A = \begin{bmatrix} 1 & 1 \\ 1 & 0.999\,999\,999\,9 \end{bmatrix}$$

and

$$A^{-1} = \begin{bmatrix} -999\,999\,999\,9 & 10^{10} \\ 10^{10} & -10^{10} \end{bmatrix}$$

the condition number, in l_∞-norm, is given by

$$\text{cond}\,(A) = \|A\|\,\|A^{-1}\| = 4 \times 10^{10}$$

This large value of cond (A) means that A must be very close to being singular, i.e. almost equal in norm to some singular matrix. In fact, by choosing a certain ΔA as

$$\Delta A = \begin{bmatrix} 0 & -5 \times 10^{-11} \\ 0 & 5 \times 10^{-11} \end{bmatrix}$$

we get

$$A + \Delta A = \begin{bmatrix} 1 & 0.999\,999\,999\,95 \\ 1 & 0.999\,999\,999\,95 \end{bmatrix}$$

which is singular. Despite its very small norm ($\|\Delta A\|_\infty = 5 \times 10^{-11}$), this increment (perturbation) ΔA transformed A into a singular matrix. This means that A must have been originally very close to being singular. This can be interpreted by stating that since cond (A) is so large, the relative difference between A and the singular matrix closest to it is very small. In fact, a measure of this closeness is given by $1/\|A^{-1}\|$ for when A is singular, $\|A^{-1}\|$ becomes infinite, i.e.

$$\frac{1}{\|A^{-1}\|} = \min\,(\|A - S\|)$$

the minimum being taken over all the singular matrices S. Hence

$$\frac{1}{\text{cond}\,(A)} = \min\,(\|A - S\|/\|A\|)\,,$$

a result which appeared in Kahan (1966). For the above example, in l_∞-norm,

$$\|A^{-1}\| = 2 \times 10^{10}$$

and

$$\frac{1}{\|A^{-1}\|} = 5 \times 10^{-11} \quad (= \|\Delta A\|)$$

This latter value $(1/\|A^{-1}\|)$ is in fact the key to a suitable choice of the matrix ΔA, as shown above.

The two diagrams below are intended to visualize and clarify the concept of nearness to singularity. Around every matrix A (represented by a point in space) there exists a region of radius $\|\Delta A\|$ wherein any matrix is practically indistinguishable from A in terms of data uncertainties or round-off errors. Figure 1 shows the configuration of a well-conditioned matrix, located at a relatively large distance from the nearest singular matrix. In Fig. 2, the matrix A is ill-conditioned since the spherical region of radius $\|\Delta A\|$ encloses some singular matrices.

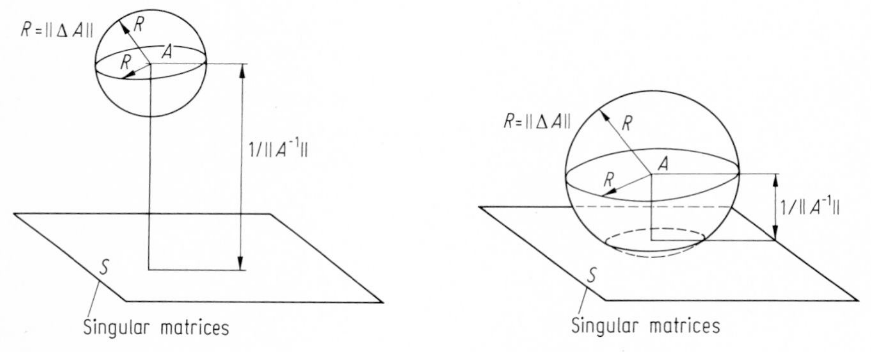

Fig. 1 **Fig. 2**

In the next section, we will see how $\|\Delta A\| \, \|A^{-1}\|$ is involved in the evaluation of error bounds in problems such as matrix inversion and sensitivity of solutions of simultaneous linear equations. For the time being we can state that systems with a small value of $\|\Delta A\| \, \|A^{-1}\|$, i.e. for which

$$\|\Delta A\| \ll 1/\|A^{-1}\|$$

are less susceptible to perturbation-induced solution errors. Instead if $1/\|A^{-1}\| < \|\Delta A\|$ or that $1/\text{cond}\,(A) <$ machine precision w.r.t. $\|A\|$ under rounding errors, the matrix A becomes indistinguishable from a singular matrix, rendering the computation meaningless. In the LINPACK user's guide (see Dongarra et al. (1979)), A is considered *singular to working precision* if the logical expression 1.0 + RCOND.EQ.1.0 is true (RCOND is the reciprocal of cond).

It becomes apparent that $\|A^{-1}\|$ is the most important part of cond (A); the one causing greater calculation difficulties due to rounding errors. Usually, $\|A^{-1}\|$ is calculated through solving the linear equations

$$Ax = y \, ,$$

with y chosen so as to enhance relatively the growth of x, yielding

$$\|A^{-1}\| \cong \frac{\|x\|}{\|y\|}$$

simply because

$$\|A^{-1}\| = \sup_y \frac{\|A^{-1}y\|}{\|y\|}$$

Henceforth, cond (A) can be considered a byproduct of the solution of linear simultaneous equations. Its accuracy depends to a great extent on the algorithm used and the pivotal strategy followed. For instance, Cline, Moler, Stewart and Wilkinson (1979) proposed complete pivoting together with an LU decomposition of A for better stability. Their method has been largely implemented in the LINPACK package. O'Leary (1980) provided a simpler modification to improve the accuracy of the condition number estimate, see also Hager (1984) in this respect. Estimating condition numbers for sparse matrices is found in Grimes and Lewis (1981).

At this stage, an important fact should be pointed out. All ill-conditioned matrices being very near to a singular one, one might think that det(A) equals approximately zero or — equivalently — that an eigenvalue of A is near the origin. This is misleading, for in the case of a matrix like

$$A = \begin{bmatrix} 10^{10} & 0 \\ 0 & 10^{-10} \end{bmatrix}$$

cond $(A) = 10^{20}$, and yet det $(A) = 1$. Then again, A needs only a mere

$$\Delta A = \begin{bmatrix} 0 & 0 \\ 0 & -10^{-10} \end{bmatrix}$$

to become singular. Likewise, for the matrix given in Moler (1978), as

$$A = \begin{bmatrix} -149 & -50 & -154 \\ 537 & 180 & 546 \\ -27 & -9 & -25 \end{bmatrix}$$

which is also ill-conditioned, the eigenvalues are $\lambda = 3;\ 2;\ 1$ and det $(A) = 6$. The nearest singular matrix is

$$S = \begin{bmatrix} -149.000\,6.. & -49.998\,07.. & -154.000\,04.. \\ 536.999\,8.. & 180.000\,63.. & 545.999\,98.. \\ -27.000\,61.. & -8.998\,03.. & -25.000\,04.. \end{bmatrix}$$

Hence, neither det (A) nor its eigenvalues are a measure of ill-conditioning. This is due to the fact that

$$\operatorname{cond}(A) = \sup_{\|x\|=\|y\|=1} \frac{\|Ax\|}{\|Ay\|} \geq \frac{\max_i |\lambda_i|}{\min_i |\lambda_i|}$$

while it is not necessary that cond (A) be small for the right hand side of the inequality to remain small. Actually, it is the set of the singular values of A that gives such a measure of ill-conditioning, as the *spectral condition number* can be evaluated as

$$\operatorname{cond}(A) = \frac{\max_i \sigma_i}{\min_i \sigma_i}$$

where σ stands for singular value. For the above matrix A, we have

$$\sigma_1 = 817.760$$
$$\sigma_2 = 2.47497$$
$$\sigma_3 = 0.0029645$$

Whence

$$\operatorname{cond}(A) = 275850.9023$$

The definition of cond (A) discussed in the foregoing pages is the most popular. Other less common ones can be found in the literature. For instance, Kuperman (1971, p. 10) proposed the definition

$$\operatorname{cond}(A) = \left(\sum_{i=1}^{n} \sum_{j=1}^{n} \left| \frac{\partial \det(A)}{\partial a_{ij}} \right| \varepsilon_{ij} \right) \Big/ |\det(A)|$$

where ε_{ij} is the uncertainty in a_{ij}. When defined in this way, a condition number of less than unity would signify that the change in det (A) — represented by the numerator — does not approach det (A) in value. Hence a small uncertainty ε_{ij} would not bring det $(A + \Delta A(\varepsilon_{ij}))$ to zero under small ε_{ij}. The matrix A is then not *critically ill-conditioned.* Similar determinantal criteria, as applied to nearly-singular matrices, are also found in Noble (1969, ch. 8).

At this stage, before proceeding to the following section, it is interesting to outline a numerical application of the condition number related to the concepts of conditioning and nearness to singularity. Let A be a perfectly singular matrix. We now seek to solve the system of equations

$$Ax = 0$$

i.e. to compute a null vector of A. An excellent approximation to a null vector could be calculated using a procedure outlined by Moler (1978), termed *Inverse-Iteration Method*. It proceeds as follows:

1. Suppose A is nonsingular
2. Choose at random a vector y
3. Solve the system $Ax = y$

(Theoretically, some component(s) of x must be infinite. However, in practice — due to rounding — x will acquire a very large value but will probably not become infinite).

4. We proceed to normalize x
5. We obtain the solution as $x/\|x\|$

This simple technique is based on the equation

$$A(x/\|x\|) = y/\|x\| \cong 0$$

To visualize this, we borrow the author's illustrative example

$$A = \begin{bmatrix} 3 & 6 & 9 \\ 2 & 5 & 8 \\ 1 & 4 & 7 \end{bmatrix}$$

Choosing the vector y as

$$y^T = (3.14, 0.00, 2.72)$$

the solution vector x is found by elimination, that is

$$\left[\begin{array}{ccc|c} 3 & 6 & 9 & 3.14 \\ 0 & 2.002 & 4.003 & 1.67 \\ 0 & 0 & 5 \times 10^{-3} & -2.92 \end{array}\right]$$

yielding

$$x^T = (-5.83 \times 10^3, \quad 11.69 \times 10^3, \quad -5.85 \times 10^3)$$

and

$$\frac{x^T}{\|x\|} = (1, \quad -2.005, 1.003)$$

which is equal to the null vector within machine accuracy. The reader is also referred to Kramarz (1981) for similar ideas; regarding modification of a singular matrix into a nonsingular one by perturbation, with application to computing a null vector.

1.4 A-Priori Bounds

This section will only discuss determined systems. Perturbation analysis of indeterminate and overdetermined systems makes use of the sensitivity analysis of A^+, which relies in turn on the sensitivity of the eigenvalue problem. The analysis of these systems will thus be postponed to a later chapter. To carry out a sensitivity analysis of $Ax = b$, given that $\det(A) \neq 0$, we proceed with a study of the effects of variations in both A and b on the solution x. In other words, if both A and b undergo a perturbation of the form $A + \Delta A$ and $b + \Delta b$, how would this affect x? As seen before, ΔA and Δb could be either induced perturbations or rounding errors unavoidable in numerical computations.

Such a study falls in the category of a-priori analysis, in contrast with a-posteriori analysis (discussed in Chap. 2). It was Wilkinson who, in 1959, pioneered rounding errors' analysis in linear systems with a mammoth work on the subject. Later on, he further published a series of papers and books summarizing his work (1960, 1961, 1963, 1965, 1971). However, the idea of rounding-off errors runs still older. We could trace the concept of matrix conditioning at least back to Turing (1948). It had already figured implicitly in Von Neumann and Goldstine's paper (1947).

To see how the variations in the condition number affect the solution x, write

$$x + \Delta x = (A + \Delta A)^{-1}(b + \Delta b)$$

Now expand $(A + \Delta A)^{-1}$ to become

$$(A + \Delta A)^{-1} = A^{-1} - A^{-1}\Delta A A^{-1} + A^{-1}\Delta A A^{-1}\Delta A A^{-1} - \dots,$$

if $\|A^{-1}\Delta A\|$ is less than unity, or more rigorously if and only if the spectral radius of $A^{-1}\Delta A$ is less than unity. We can obtain, upon neglecting second order terms:

$$\Delta x \cong A^{-1}\Delta b - A^{-1}\Delta A x$$

that is

$$\frac{\|\Delta x\|}{\|x\|} \leq \|A^{-1}\| \left(\frac{\|\Delta b\|}{\|x\|} + \|\Delta A\| \right)$$

but

$$\|x\| = \|A^{-1}b\| \geqq \|A\|^{-1}\|b\|$$

from which one finally obtains

$$\frac{\|\Delta x\|}{\|x\|} \leq \|A^{-1}\|\,\|A\| \left(\frac{\|\Delta A\|}{\|A\|} + \frac{\|\Delta b\|}{\|b\|} \right)$$

$$= \text{cond}(A) \left(\frac{\|\Delta A\|}{\|A\|} + \frac{\|\Delta b\|}{\|b\|} \right)$$

which is the well known error bound. The condition number therefore determines the susceptibility of the solution of a linear system to given changes in the data, coinciding with the definition set forth by Rice (1966), see also Geurts (1982).

One could obtain the same result above by introducing the factor ε-measuring the strength of perturbation — in the equation

$$(A + \varepsilon A_1)(x + \varepsilon x_{(1)} + \dots) = b + \varepsilon b_1$$

The bracket $(x + \varepsilon x_{(1)} + \dots)$ is a series in ε made convergent by choosing $|\varepsilon|$ small enough. By regrouping terms containing ε, and setting

$$\|Ax_{(1)}\| \geqq \|A^{-1}\|^{-1} \|x_{(1)}\|$$

and $\|b_1 - A_1 x\| \leqq \|b_1\| + \|A_1\| \|x\|$

one could wind up with the same bound. Note that the condition

$$\|A^{-1} \Delta A\| < 1$$

still holds, since ε is a fictitious factor contained in ΔA. Furthermore, $(x + \varepsilon x_{(1)} + \dots)$ is a convergent series, since it is the product of the expression $(b + \Delta b)$ times the convergent series

$$A^{-1} - A^{-1} \Delta A \, A^{-1} + \dots .$$

In order to put the emphasis on first order perturbations, we prefer to include only the terms with ε. A_1 will then be of the same order of magnitude as A. Grouping the terms containing ε in the foreseen equation will therefore implicitly guarantee convergence. Note that, in the case where the term $\Delta A \, \Delta x$ is not neglected, a further minor refinement can be brought to the above set bound, e.g. the one described by Franklin (1968)

$$\frac{\|\Delta x\|}{\|x\|} \leq \frac{\text{cond }(A)}{1 - \text{cond }(A) \dfrac{\|\Delta A\|}{\|A\|}} \left(\frac{\|\Delta A\|}{\|A\|} + \frac{\|\Delta b\|}{\|b\|} \right)$$

Though very cumbersome to use, this bound does not usually alter the previous result noticeably. In fact, as Forsythe and Moler (1967) observed rather easily, when b is held constant, that

$$\frac{\|\Delta x\|}{\|x + \Delta x\|} \leqq \|A^{-1} \Delta A\|$$

$$\leqq \text{cond}(A) \frac{\|\Delta A\|}{\|A\|}$$

the inequality holding without any approximations (see exercise 1.19). In any case, cond (A) always appears in the expressions for error bounds related to linear equations; it also appears in expressions for similar bounds associated with the relative variation in A^{-1}, such as

$$\frac{\|(A+\Delta A)^{-1}-A^{-1}\|}{\|A^{-1}\|} \leq \frac{\|A^{-1}\Delta A\|}{1-\|A^{-1}\Delta A\|} \leq \frac{\operatorname{cond}(A)\dfrac{\|\Delta A\|}{\|A\|}}{1-\operatorname{cond}(A)\dfrac{\|\Delta A\|}{\|A\|}}$$

which is identical to that obtained for $\|\Delta x\|/\|x\|$. This should not come as a surprise, as computation of A^{-1} is equivalent to solution of the equation $AX = B$ with $B = I$. Bounds associated with A^{-1} are further mentioned in exercise 1.18.

For first order perturbations, the reader should further note, the relative variation in $\|A^{-1}\|$ is approximated by

$$\|\Delta A\| \cdot \|A^{-1}\| = \|\Delta A\|/(1/\|A^{-1}\|) = \frac{\text{radius of sphere}}{\text{distance to a singular matrix}}$$

(refer to Fig. 1 and 2, Sect. 1.3)

Therefore, the more accentuated the ill-conditioning of A, the greater the susceptibility of A^{-1} to rounding- or perturbation-induced errors.

The calculations involved in applying the above bounds for either of $Ax = b$ or A^{-1} are short and straightforward. Even for $\|A^{-1}\|$, we need only solve a set of linear equations

$$Ax = y \qquad \text{(see Sect. 1.3)}$$

Beside being referred to as simple, elegant and easy to work with, these bounds are also given first priority when the sensitivity analysis of a linear system is performed. Moreover, in virtue of the bounds' simplicity, some perturbations occasionally cause inequalities to become equations (see van der Sluis (1970a)). And for each inequality, there exist matrices for which the equality holds.

To exemplify the above results in a simple way, let us consider the two equations seen in Sect. 1.2, namely

$$A = \begin{bmatrix} 2 & 1 \\ 1 & 2 \end{bmatrix}, \quad b = \begin{bmatrix} 2 \\ -1 \end{bmatrix}, \quad \Delta A = \begin{bmatrix} 0 & 0 \\ 0.01 & -0.02 \end{bmatrix}$$

The results obtained on a 10-digit machine — to minimize the effect of rounding compared to the induced ΔA — were as follows:

$$x = (1.666666667, -1.333333333)^T$$
$$x + \Delta x = (1.681355932, -1.362711864)^T$$

Then, using the l_1-norm

$$\frac{\|\Delta x\|}{\|x+\Delta x\|} = \frac{0.044067796}{3.044067796} = 0.014476615$$

which represents an error of 1.5% approximately. Then with

$$A^{-1}\,\Delta A = \begin{bmatrix} -0.003\,333\,333 & 0.006\,666\,667 \\ 0.006\,666\,667 & -0.013\,333\,333 \end{bmatrix}$$

we get

$$\frac{\|\Delta x\|}{\|x + \Delta x\|} \leqq \|A^{-1}\,\Delta A\| = 0.01999$$

which in turn represents an error of approximately 2%, thus setting a bound larger than the true error. To compare the magnitude of this perturbation in the solution with that incurred had A been ill-conditioned, let us consider the example in Sect. 1.1, namely

$$A = \begin{bmatrix} 1 & 1 \\ 0.49 & 0.51 \end{bmatrix}, \qquad b = \begin{bmatrix} 2 \\ 1 \end{bmatrix}, \qquad \Delta A = \begin{bmatrix} 0 & 0 \\ 0.005 & -0.01 \end{bmatrix}$$

Here the value of $\|\Delta A\|/\|A\|$ is the same as above. Yet since cond (A) is large, a large error should be expected. In fact

$$x = (1,\ 1)^T$$

$$x + \Delta x = (0,\ 2)^T$$

that is

$$\frac{\|\Delta x\|}{\|x + \Delta x\|} = \frac{1 + 1}{2} = 1$$

i.e. quite a huge one hundred percent error has been induced. Alternatively

$$A^{-1}\,\Delta A = \begin{bmatrix} -0.25 & 0.5 \\ 0.25 & -0.5 \end{bmatrix}$$

so that

$$\frac{\|\Delta x\|}{\|x + \Delta x\|} \leqq \|A^{-1}\,\Delta A\|$$
$$= 1$$

Seen the influence of cond (A) on the perturbed solution, the question arises as to whether cond (A) could be reduced, or whether the errors induced in general could be reduced. This question has been the major concern of many a numerical analyst for more than two decades. Answering it always relied on a combination of adequate pivoting along with scaling or equilibration.

1.5 Pivoting and Equilibration

To demonstrate the effect of a good pivoting strategy in improving the solution precisionwise, we borrow an example treated by Wilkinson (1965, p. 216). Solve

$$\begin{bmatrix} 0.000\,003 & 0.213\,472 & 0.332\,147 \\ 0.215\,512 & 0.375\,623 & 0.476\,625 \\ 0.173\,257 & 0.663\,257 & 0.625\,675 \end{bmatrix} \begin{bmatrix} x_1 \\ x_2 \\ x_3 \end{bmatrix} = \begin{bmatrix} 0.235\,262 \\ 0.127\,653 \\ 0.285\,321 \end{bmatrix}$$

with a 6-decimal-digit machine. If 0.000003 is to be taken as a pivot, the reduced system of equations becomes

$$\begin{bmatrix} 0.000\,003 & 0.213\,472 & 0.322\,147 \\ 0 & -15\,334.9 & -23\,860.0 \\ 0 & -12\,327.8 & -19\,181.7 \end{bmatrix} \begin{bmatrix} x_1 \\ x_2 \\ x_3 \end{bmatrix} = \begin{bmatrix} 0.235\,262 \\ -16\,900.5 \\ -13\,586.6 \end{bmatrix}$$

Taking -15334.9 as a pivot for another eliminatory step, we obtain

$$\begin{bmatrix} 0.000\,003 & 0.213\,472 & 0.322\,147 \\ 0 & -15\,334.9 & -23\,860.0 \\ 0 & 0 & -0.500\,000 \end{bmatrix} \begin{bmatrix} x_1 \\ x_2 \\ x_3 \end{bmatrix} = \begin{bmatrix} 0.235\,262 \\ -16\,900.5 \\ -0.200\,000 \end{bmatrix}$$

Using back-substitution, we obtain

$$\hat{x}_3 = 0.400000$$
$$\hat{x}_2 = 0.479723$$
$$\hat{x}_1 = -1.33333$$

To show how poor these results are, Wilkinson compared them with those obtained when a better pivoting strategy is adopted. Going back to the initial equations, Wilkinson selected 0.215512 as the pivot, instead of 0.000003. He obtained

$$\begin{bmatrix} 0.215\,512 & 0.375\,623 & 0.476\,625 \\ 0 & 0.361\,282 & 0.242\,501 \\ 0 & 0 & 0.188\,856 \end{bmatrix} \begin{bmatrix} x_1 \\ x_2 \\ x_3 \end{bmatrix} = \begin{bmatrix} 0.127\,653 \\ 0.182697 \\ 0.127\,312 \end{bmatrix}$$

And upon back-substitution, the solution comes as

$$\hat{x}_3 = 0.674122$$
$$\hat{x}_2 = 0.0532050$$
$$\hat{x}_1 = -0.991291$$

The correct answer, down to ten decimal digits, comes as

$$
\begin{aligned}
x_3 &= 0.6741214694 \\
x_2 &= 0.05320393391 \\
x_1 &= -0.9912894252
\end{aligned}
$$

Comparison elicits how a good pivoting strategy brings about remarkable accuracy improvements. A good pivotal strategy is therefore one which allows for interchanging rows with poor pivots with ones having good pivots using elementary row operations.

One might argue that pivoting should not improve the solution, since the product of the pivots equals det (A), the value of which is independent of the pivoting strategy or sequence. Therefore, if good pivots are used at the start, the poor ones left will be used near the end of the process, jeopardizing the attained accuracy. This reasoning is fallacious because, as Wilkinson pointed out, errors propagate and amplify during calculations, thence the earlier their elimination, the lesser their importance near the end of the calculations. This is visible in the foregoing example, wherewith starting with the smaller pivot did not make the last pivot any larger. On the contrary it usually introduces a very large pivot in the next stage, returning to normal in the last stage but one. The matrix then becomes disequilibrated, this situation affecting greatly the solution as will be seen later.

Scaling of the equations $Ax = b$ is in fact mainly a scaling of the vectors of A, to make them more or less of the same norm. The obtained system of equations is henceforth said to be in an *equilibrated form*. For example,

$$
A = \begin{bmatrix} 2 & 1 & 3 \times 10^6 \\ 1 & -1 & 10^6 \\ 1 & -2 & 0 \end{bmatrix}
$$

can be so scaled to obtain

$$
A' = \begin{bmatrix} 0.2 & 0.1 & 0.3 \\ 0.1 & -0.1 & 0.1 \\ 0.1 & -0.2 & 0 \end{bmatrix}
$$

This procedure can invariably be carried out for any matrix A (cf. Forsythe and Moler (1967)). However, the equilibration mode of a matrix is not unique, whence the possibility of following different pivotal strategies when using Gauss elimination. Equilibration is nonetheless advantageous, as it generally decreases the condition number of a matrix. Van der Sluis (1969) showed that for different norms in current use, the condition number can be minimized by row scaling down to unity norm. According to Wilkinson (1965, p. 192), a matrix with unity length rows and columns is termed *equilibrated*. For instance, unitary matrices are readily equilibrated, and correspondingly possess stable numerical operations.

We will reproduce hereafter the simple result derived by Dahl (1978) to visualize how scaling reduces the condition number, thus making the system $Ax = b$ less susceptible to perturbation errors. Dahl obtained that

$$
\frac{\|\Delta y\|}{\|y + \Delta y\|} \leqq \max_{i,j} |\Delta a'_{ij}| \, \|A^{-1}\| \, \|D_1^{-1} \{1\} D_2^{-1}\|
$$

Here, y is the solution for the new scaled system obtained by respectively premultiplying then postmultiplying $Ax = b$ by the two diagonal matrices D_1 and D_2. $\max_{i,j} |\Delta a'_{ij}|$ is a bound on the perturbations of the elements of the scaled system and $\{1\}$ is a matrix the elements of which are all unity. To see how easily this result follows from $Ax = b$, we write

$$\begin{aligned} D_1 b &= D_1(A + \Delta A)\, D_2 D_2^{-1}(x + \Delta x) \\ &= (A' + \Delta A')\,(y + \Delta y) \end{aligned}$$

with

$$\begin{aligned} D_1 A D_2 &= A' \\ D_1\, \Delta A\, D_2 &= \Delta A' \end{aligned}$$

Naturally, our interest focuses next on evaluating

$$\|\Delta y\| / \|y + \Delta y\|$$

Applying the bound mentioned a few pages ago, we get

$$\begin{aligned} \frac{\|\Delta y\|}{\|y + \Delta y\|} &\leqq \|\Delta A'\|\ \|A'^{-1}\| \\ &\leqq k(n, t) \cdot a\ \|A^{-1}\|\ \|D_1^{-1}\{1\}\ D_2^{-1}\| \end{aligned}$$

Here $\|\Delta A'\| \leqq k(n, t) \cdot a\ \|\{1\}\|$, where $k(n, t)$ represents the machine accuracy with respect to A and a is the maximum among the absolute values of the elements of A'. The effect of scaling was elicited by the author through considering

$$A = \begin{bmatrix} 1/2 & 1/3 & 1/4 \\ 1/3 & 1/4 & 1/5 \\ 1/4 & 1/5 & 1/6 \end{bmatrix}$$

of which the inverse A^{-1} comes as

$$A^{-1} = \begin{bmatrix} 72 & -240 & 180 \\ -240 & 900 & -720 \\ 180 & -720 & 600 \end{bmatrix}$$

For the unscaled system, relative to an l_∞-norm, we obtain

$$a\|A^{-1}\|\ \|\{1\}\| = {}^1/_2\,(240 + 900 + 720)\,(3) = 2790$$

As for the scaled system, he chose

$$D_1 = \operatorname{diag}(1, 3/2, 2)\,, \qquad D_2 = I$$

thus obtaining for the same norm

$$a\|(D_1A)^{-1}\|\ \|\{1\}\| = {}^1/_2\,(240 + 900(2/3) + 720({}^1/_2))\,(3) = 1800$$

For further clarification of this result, let us consider the solution of the system $Ax = b$; A being the same as above, and b being defined by

$$b = (1, 1, 1)^T$$

we obtain, on an HP-41CV machine, a solution

$$\hat{x} = (11.99999852, -59.99999434, 59.99999539)^T$$

Simultaneously, solving the system

$$D_1Ax = D_1b$$

with D_1 set as in the example above would yield:

$$\hat{x} = (12.00000001, -60.00000012, 60.00000014)^T$$

The exact solution for this system comes as

$$x = (12, -60, 60)^T$$

whence, upon comparison, we can assess the importance of the improvement scaling brings about in the solution.

The matrix $D_1 = \text{diag}\,(1, 3/2, 2)$ is of course not the best row scaling matrix. Choosing D_1 is generally not easy, especially since D_1 has no one unique definition. A configuration of D_1 that decreases $\|A^{-1}\,D_1^{-1}\|$ usually augments $\|D_1\,\Delta A\|$. van der Sluis (1970b) suggested the choice according to

$$d_{ii} = \max_k \min_j \left|\frac{a_{kj}}{a_{ij}}\right|$$

which would prevent any row of A being dominated by another. Such a choice, while not risking to worsen a situation, still runs the risk of not improving it. For the previous example, with $D_1 = \text{diag}\,(1, 5/4, 3/2)$ according to the above definition, the scaled equations become

$$\begin{bmatrix} 1/2 & 1/3 & 1/4 \\ 5/12 & 5/16 & 1/4 \\ 3/8 & 3/10 & 1/4 \end{bmatrix} \begin{bmatrix} x_1 \\ x_2 \\ x_3 \end{bmatrix} = \begin{bmatrix} 1 \\ 5/4 \\ 3/2 \end{bmatrix}$$

the solution of which is

$$\hat{x} = (11.99999950, -59.99999813, 59.99999850)^T$$

This result is not of an improvement over the one obtained previously, but is slightly better than that obtained in the first place.

An interesting problem would indeed concern the minimization of cond $(D_1 A D_2)$ for D_1 and D_2. Unfortunately, the results obtained in this direction are not practical. But this is hardly discouraging as Golub and Van Loan (1983, p. 73) truly remarked, for already the bounds are heuristic and it makes little sense to minimize exactly a heuristic bound. Instead, it is quite sufficient to find an approximate but fast and practical way to improve the quality of the computed $\hat{x}$.

Usually, the system $Ax = b$ is only row-scaled, that is $D_2 = I$ like in the previous example. Column scaling will only lead to scaling the solution vector itself. van der Sluis (1970b) noted that pivoting and row equilibration exert rather independently their action on the algorithm's stability. The reader interested in practical effects of various pivoting and equilibration strategies is referred to work by Curtis and Reid (1972).

On the other hand, Householder (1964) suggested a method for preventing error propagation during Gauss elimination, by fixing the condition number of A so that it doesn't grow larger. This is achieved by choosing the elementary operations R^i, used in succession to bring A to an upper triangular form, as unitary (cf. exercise 1.4). Householder's famous transformation is given by

$$R^i = I - 2w^i w^{i*}$$

with

$$w^{i*} w^i = 1$$

It is both unitary and Hermitian.

Still, the foregoing a-priori bounds — as first derived by Wilkinson — are not judged fully satisfactory by scholars. Wilkinson himself confessed that they were by no means the best possible. He realized that they could still be narrowed further by elaborate argumentation (Wilkinson, (1961)), since in that form they did not account for statistical effects. Furthermore, the matrix A itself depends on the transformation of A into a sequence of matrices A^1, A^2, ... until a triangular form (like for example in Gauss elimination) is obtained. Chartres and Geuder (1967) derived bounds which can be evaluated during the computation itself. Using information generated during the solution process, they could avoid error bound magnitude exaggeration, as is the case with Wilkinson's a-priori analysis. They applied this method in relation to LU factorization. Unfortunately, the bounds thus obtained depend on the particular sequence of back-substitutions. Derivations of similar nature can also be found in work by Loizou (1968).

It was only in 1979, with the publication of Skeel's work, that the concept of conditioning was questioned. Like Hamming (1971), Skeel argues that the term *ill-conditioned system* is itself ill-defined. It is rather vague to simply state that small changes in the initial system can produce large alterations of the result. The term "relatively small — or large — change" should rather be used, if we are to take floating-point arithmetic as a serious matter. Further, it makes no difference how, or using which method, we scale A to make the answer insensitive to small changes in the original coefficients, for in every such case the system will remain ill-conditioned. We should in fact try to scale the whole system, including the vector b. For this we need a new definition of the condition number that would relate errors more correctly.

The main difference between the first approach and Skeel's is easily grasped. For instance, instead of using the form $\|\Delta A\|/\|A\|$ which does not account for the relative error in each coefficient, one should use $\| |A^{-1}| |\Delta A| \|$, where $|A|$ is the matrix of absolute values of matrix A. Beside appearing more logical, such a practice brought great improvements in the understanding of error propagation phenomena.

To obtain an error bound for $\|\Delta x\|/\|x\|$ in terms of the variations in both A and b, we write as before

$$(A + \Delta A)(x + \Delta x) = b + \Delta b$$

or

$$\Delta x = -A^{-1} \Delta A(x + \Delta x) + A^{-1} \Delta b$$

so that

$$|\Delta x| \leqq (|A^{-1}| \, \Delta A \, |x| + |A^{-1}| \, \Delta b|) + |A^{-1}| \, \Delta A \, |\Delta x|$$

It follows directly that

$$\frac{\|\Delta x\|}{\|x\|} \leq \frac{\| |A^{-1}| \, |\Delta A| \, |x| + |A^{-1}| \, |\Delta b| \|}{(1 - \| |A^{-1}| \, |\Delta A| \|) \, \|x\|}$$

a result which represents an alternative to Wilkinson's bounds. As for the condition number, it is given by

$$\lim_{\varepsilon(\Delta A, \Delta b) \to 0} \frac{\|\Delta x\|/\|x\|}{\varepsilon(\Delta A, \Delta b)}$$

where $\varepsilon(\Delta A, \Delta b)$ represents the relative error in the data. Note that the above definition does not differ from the previous one, since for

$$\|\Delta x\|/\|x\| \leqq \|A\| \, \|A^{-1}\| \, \frac{\|\Delta A\|}{\|A\|}$$

ε was represented by $\|\Delta A\|/\|A\|$, and $\text{cond}(A) = \|A\| \, \|A^{-1}\|$. In the present case, we simply take

$$|\Delta A| \leqq \varepsilon |A| \quad \text{and} \quad |\Delta b| \leqq \varepsilon |b|$$

thence obtaining directly

$$\text{condition number} = \frac{\| |A^{-1}| |A| |x| + |A^{-1}| |b| \|}{\|x\|}$$

which is a condition number for the whole system $Ax = b$; in contrast with the expression for cond (A) set forth previously. Seemingly, Skeel realized that to match both definitions — in cases where variations in A and b occur — one should implicitly use two different numbers ε, namely ε_1 for ΔA and ε_2 for Δb. Unless he did this, then for the old bound

$$\|\Delta x\|/\|x\| \leqq \|A\| \|A^{-1}\| \{(\|\Delta A\|/\|A\|) + (\|\Delta b\|/\|b\|)\}$$

one would get

$$\text{cond}(A) = 2\|A\| \|A^{-1}\|$$

To overcome this difficulty, Skeel associated one condition number to the variation in A, namely

$$\| |A^{-1}| |A| |x| \|/\|x\|$$

and another to variations in b, in turn

$$\| |A^{-1}| |b| \|/\|x\| = \| |A^{-1}| |Ax| \|/\|x\|$$

As both numbers come to nearly coincide, Skeel concluded that there existed one condition number that could account for both kinds of perturbations; a condition number of the solution, or more simply of A, given by

$$\text{cond}(A) = \| |A^{-1}| |A| \|$$

This expression contrasts with the previous $\|A^{-1}\| \|A\|$.

It is however very unfortunate that Skeel's definition of cond (A), despite its being more realistic as a sensitivity criterion, suffers from an impossibility of reducing it through scaling. This is easily seen if we consider the fact that for any diagonal matrix D

$$\text{cond}(DA) = \| |A^{-1}D^{-1}| |DA| \| = \text{cond}(A)$$

Row scaling must accordingly have no effect on accuracy, a statement which is in fact all but true. This should explain why no optimum way for scaling a matrix exists.

On the other hand, cond $(A) = \|A^{-1}\| \|A\|$ allows for scaling to improve accuracy, a fact well confirmed by practical examples. It also allows the use of unitary operations to fix the condition number so that it does not grow larger during computation, as in Householder's transformation. An interesting question is now whether such manipulations could be possible for the quantity cond $(A) = \| |A^{-1}| |A| \|$.

To compare both Skeel's and Wilkinson's bounds, we will hereafter use them in conjunction with the example presented by Hamming (1971), namely

$$A = \begin{bmatrix} 3 & 2 & 1 \\ 2 & 2\varepsilon & 2\varepsilon \\ 1 & 2\varepsilon & -\varepsilon \end{bmatrix}, \qquad b = \begin{bmatrix} 3 + 3\varepsilon \\ 6\varepsilon \\ 2\varepsilon \end{bmatrix}$$

of which the exact solution comes as

$$x = (\varepsilon, 1, 1)^T$$

According to the definition of cond (A) as $\|A^{-1}\| \, \|A\|$, the matrix A would be ill-conditioned for small values of $|\varepsilon| (\|A^{-1}\| \, \|A\| = 36 \times 10^5$ for $|\varepsilon| = 10^{-6})$. However, for the particular choice of b depicted above, the system is well-conditioned according to Skeel's definition. Indeed

$$A^{-1} = \frac{1}{1 - 1.8\varepsilon} \begin{bmatrix} -0.6\varepsilon & 0.4 & 0.2 \\ 0.4 & \dfrac{-0.1}{\varepsilon} - 0.3 & \dfrac{0.2}{\varepsilon} - 0.6 \\ 0.2 & \dfrac{0.2}{\varepsilon} - 0.6 & \dfrac{-0.4}{\varepsilon} - 0.6 \end{bmatrix}$$

whence

$$|A^{-1}| \, |A| = \frac{1}{1 - 1.8\varepsilon} \begin{bmatrix} 1 + 1.8\varepsilon & 2.4\,\varepsilon & 1.6\varepsilon \\ \dfrac{0.4}{\varepsilon} + 1.2 & 1.4 - 0.6\varepsilon & 0.8 \\ \dfrac{0.8}{\varepsilon} & 1.6 & 1 - 0.6\varepsilon \end{bmatrix}$$

and

$$\text{cond}\,(A, x) = \frac{\| |A^{-1}| \, |A| \, |x| + |A^{-1}| \, |b| \|}{\|x\|} = \frac{6 - 2.4\varepsilon}{1 - 1.8\varepsilon}$$

which shows that the system is well-conditioned. However

$$\text{cond}\,(A) = \| |A^{-1}| \, |A| \| = \frac{0.8\varepsilon^{-1} + 2.6 - 0.6\varepsilon}{1 - 1.8\varepsilon}$$

indicating that for small values of $|\varepsilon|$, the system would be ill-conditioned for some other right-hand side term b. On the contrary, the system with the matrix A as

$$A = \begin{bmatrix} 3 & 2 & 1 \\ 2\varepsilon & 2\varepsilon & 2\varepsilon \\ \varepsilon & 2\varepsilon & -\varepsilon \end{bmatrix}$$

is well-conditioned whatever the choice of b may be, since in this case

$$\| |A^{-1}| \, |A| \| = 17/2$$

and

$$\text{cond}\,(A, x) \leqq 2 \| \, |A^{-1}| \, |A| \, \| = 17$$

Therefore, for any vector b, the solution must be expected to be only as accurate as the machine itself. Indeed, the choice of b as

$$b = (1/\varepsilon, 1, 1)^T, \qquad \varepsilon = 10^{-9}$$

and the solution of the system on an HP-15C calculator with floating-point arithmetic yields

$$\hat{x} = (-4.7333 \ldots \times 10^{-2}, \qquad 5.000000001 \times 10^8, \qquad -5.800 \ldots \times 10^{-2})^T$$

The exact solution x being given by

$$x = (0, 5 \times 10^8, 0)^T$$

we have that

$$\frac{\|\Delta x\|}{\|x\|} \cong 0.1/(5 \times 10^8) = 2 \times 10^{-10}$$

The reader may try any other vector b of his choice to investigate this accuracy. In fact, applying Skeel's bound to our example gives

$$\begin{aligned} \frac{\|\Delta x\|}{\|\hat{x}\|} &\leqq 5 \times 10^{-10} \times \frac{\| \, |A^{-1}| \, |A| \, |\hat{x}| + |A^{-1}| \, |b| \, \|}{\|\hat{x}\|} \\ &\cong 5 \times 10^{-10} \times \frac{3 \times 10^9}{5 \times 10^8} \\ &= 3 \times 10^{-9} \end{aligned}$$

being the machine's accuracy itself with respect to $\|x\|$.

To compare this result with the one obtained from Wilkinson's a-priori bound — in case we had to use cond $(A) = \|A^{-1}\| \, \|A\|$ — we obtain

$$\frac{\|\Delta x\|}{\|\hat{x}\|} \leqq \|A^{-1} \Delta A\| + 0(\|\Delta b\|)$$

with

$$\|A^{-1} \Delta A\| \leqq \text{cond}\,(A) \frac{\|\Delta A\|}{\|A\|} = 6 \times 10^{+9} \times 3 \times 10^{-9} = 18$$

a result in itself very misleading.

Comparison might be rather unfair for Hamming's example; where well- or ill-conditioning partly accounted for the particular configuration of b considered. There could be no doubt now, irrespective of the choice of b, as to how poor the a-priori bounds are, as illustrated by the second example above. This fact did not elude Wilkinson (1965, p. 190) who noticed that by substituting $\|A^{-1}\| \, \|\Delta A\|$ for $\|A^{-1} \Delta A\|$, pessimistic results would be obtained. As he noticed, for cases where $\Delta A = \alpha A$, then $\|A^{-1} \Delta A\| = \alpha$, i.e. it is independent of cond (A).

At this stage, one might wonder, since Skeel's bound sounds more realistic, why it had to be introduced by a whole section on a-priori bounds that yield pessimistic error criteria. To answer this, we must first point out that, apart from any historical sequence, Wilkinson's bounds are easier and quicker to apply. Should one start off with them, and should he find that the error is tightly bounded, he would have a guarantee — and a good enough one — of the solution's accuracy. Furthermore, the a-priori bounds provide a crude estimate of the error in the solution without indulging in the solution process itself, as might be necessary here to evaluate A^{-1}.

On the other hand, Skeel's bound is computed a-posteriori because it involves the solution $\hat{x}$. It is only in some cases where the system is well-conditioned irrespective of the right-hand side b that we have

$$\frac{\|\Delta x\|}{\|x\|} \leqq 2\varepsilon \frac{\|\,|A^{-1}|\,|A|\,\|}{1 - \varepsilon \,\|\,|A^{-1}|\,|A|\,\|}$$

serving as an a-priori bound. Even here, A^{-1} has to be used.

As a general method for investigating good- from ill-conditioning, we can compute in order the following functions:

$$a = \|A^{-1}\| \, \|A\| \, \varepsilon$$

$$b = \|\,|A^{-1}|\,|A|\,\| \, \varepsilon$$

$$c = \frac{\|\,|A^{-1}|\,|A|\,|x| + |A^{-1}|\,|b|\,\|}{\|x\|} \varepsilon$$

where ε represents either the machine's precision or the uncertainties in our data. If a is high-valued, then we can conclude that $Ax = b$ is not necessarily ill-conditioned. Next trying out b, if still high, then the value of c will answer our querry. On the other hand, a being small-valued guarantees that b and c are small.

Another important part of Skeel's work is indeed a remark concerning Gauss elimination method with partial or complete pivoting. It is interesting to note that cond (A, x) might grow larger after pivoting. For instance, in Hamming's example cond $(A, x) \cong 6$. For the system after one eliminatory step, that is when

$$A' = \begin{bmatrix} 3 & 2 & 1 \\ 0 & -4/3 + 2\varepsilon & \dfrac{-2}{3} + 2\varepsilon \\ 0 & -2/3 + 2\varepsilon & -1/3 - \varepsilon \end{bmatrix}, \qquad b' = \begin{bmatrix} 3 + 3\varepsilon \\ -2 + 4\varepsilon \\ -1 + \varepsilon \end{bmatrix}$$

$$\text{cond}\,(A', x) = \frac{0.8\varepsilon^{-1} - 3 + 3\varepsilon}{1 - 1.8\varepsilon}$$

meaning that Gauss elimination is not asymptotically stable for any pivotal strategy depending only on the coefficient matrix A. This is visualized by trying to solve $Ax = b$ in Hamming's example on an HP-15C calculator with floating-point arithmetic and a ten-digit accuracy. By taking $\varepsilon = 10^{-9}$ the results were

$$\hat{x} = (1.166\,666\,667 \times 10^{-9}, 1.024\,351\,388, 9.512\,972\,235 \times 10^{-1})^T$$

with an accuracy of

$$\frac{\|\Delta x\|}{\|x\|} \cong 5 \times 10^{-2}$$

which is very poor, notwithstanding the number of machine digits.

Skeel suggested a scaling procedure to overcome this problem remarkable in that it works perfectly in most cases. The matrix D, chosen to scale the system $Ax = b$ into $DAx = Db$, is given by

$$d_{ii} = 1 \Big/ \left(\sum_j |a_{ij}|\,|\hat{x}_j| + |b_i| \right)$$

This scaling procedure has the advantage of reducing the residual Dr of the scaled system to a value within machine accuracy (cf. Sect. 2.5). Unfortunately, it runs the risk of increasing the value of $\|A^{-1}D^{-1}\|$. For the specific example listed above, it has decreased this last quantity, thence improving greatly the accuracy. For Hamming's example, D comes as

$$D = \text{diag}\left(\frac{1}{6 + 6\varepsilon}, \frac{1}{12\varepsilon}, \frac{1}{6\varepsilon} \right)$$

Solving the scaled equations — $DAx = Db$ — on an HP-15C for a value of $\varepsilon = 10^{-9}$, we get

$$\hat{x} = (9.999\,999\,996 \times 10^{-10}, 1, 9.999\,999\,998 \times 10^{-1})^T$$

with accuracy

$$\frac{\|\Delta x\|}{\|x\|} \cong 5 \times 10^{-10}$$

which is of the order of the machine's accuracy itself.

Although Skeel's scaling criterion seems ideal, it still faces a difficulty in its application, since it needs, again, an estimate of $\hat{x}$. This could be overcome by solving first the unscaled system, then proceeding to solve the scaled one. This exercise is time consuming, but it pays off in terms of accuracy — especially with large and sparse systems.

Among this vast ocean of endeavour for tighter error bounds, it is undoubtedly Oettli and Prager (1964) who laid the foundations for a-posteriori analysis. They introduced a technique called interval analysis that has inspired Skeel's work. The chapter to come will discuss this technique in detail.

1.6 Sensitivity Analysis: a Circuit Theory Interpretation

In this section, we will discuss a problem akin to the foregoing error analysis, namely the sensitivity of x with respect to the elements of A and b. If A and b are functions of some parameters α_j; $j = 1, \ldots, m$, then to obtain $\partial x_k/\partial \alpha_j$, we simply differentiate $Ax = b$ to obtain

$$\frac{\partial A}{\partial \alpha_j} x + A \frac{\partial x}{\partial \alpha_j} = \frac{\partial b}{\partial \alpha_j}$$

whence

$$\frac{\partial x_k}{\partial \alpha_j} = e^T A^{-1} \left(\frac{\partial b}{\partial \alpha_j} - \frac{\partial A}{\partial \alpha_j} x \right)$$

where $e^T = (0, \ldots, 0, 1, 0, \ldots, 0)$ with the unity element in the kth place. In the special case where the above expression vanishes totally for a certain value of k and j, the variable x_k is said to be insensitive to perturbations in α_j. The reader interested in further reading on insensitivity of linear models and its applications is referred to work by Rosenthal (1976).

We will now consider a physical problem which, to the author's opinion, might offer a generalization of Engineers' accomplishments in sensitivity analysis, namely the definition of the *system adjoint* to $Ax = b$. We start by writing the equation $Ax = b$ in the form

$$[A \mid -I] \left[\frac{x}{b} \right] = 0$$

where $[A \mid -I]$ is equivalent to the *cut-set* matrix describing Kirchhoff's current law in an electrical network. Suppose we devise a dual or adjoint system described by $A^T y = -e$, that is

$$[I \mid A^T]\begin{bmatrix} e \\ \hdashline y \end{bmatrix} = 0$$

where e is a vector yet undefined. The solution to the above equation will necessarily take the form

$$\begin{bmatrix} e \\ \hdashline y \end{bmatrix} = \begin{bmatrix} A^T \\ \hdashline -I \end{bmatrix} z$$

z being an arbitrary vector. Again, in an electrical network, $[I \mid A^T]$ may represent the *tie-set* matrix describing Kirchhoff's voltage law. The vector $(x \mid b)^T$, $(e \mid y)^T$ and z are then respectively equivalent to the branch currents, branch voltages and nodal voltages.

We will further define the two systems S and $\hat{S}$, illustrated by the two diagrams

$Ax = b$	$A^T y = -e$
S	$\hat{S}$

It is easily seen that

$$(e^T \mid y^T)\begin{pmatrix} x \\ \hdashline b \end{pmatrix} = 0$$

This follows immediately from the fact that the last inner product is equal to $z^T(A \mid -I)\begin{pmatrix} x \\ \hdashline b \end{pmatrix}$, which is null as seen above. By further writing the relation of orthogonality as

$$e^T x + y^T b = 0$$

one can differentiate the expression with respect to the elements of A, that is a_{ij}, to obtain

$$e^T \frac{\partial x}{\partial a_{ij}} + \frac{\partial y^T}{\partial a_{ij}} b = 0$$

But from the expression $y^T A = -e^T$, one can write that $\dfrac{\partial y^T}{\partial a_{ij}}$ is given by

$$\frac{\partial y^T}{\partial a_{ij}} A + y^T \frac{\partial A}{\partial a_{ij}} = 0$$

wherefrom we can derive an expression for the sensitivity, $\dfrac{\partial x}{\partial a_{ij}}$, as

$$e^T \frac{\partial x}{\partial a_{ij}} - y^T \frac{\partial A}{\partial a_{ij}} A^{-1} b = 0$$

or

$$e^T \frac{\partial x}{\partial a_{ij}} = y^T \frac{\partial A}{\partial a_{ij}} x$$

This is exactly what engineers arrive at in calculating sensitivities, for to compute $\partial x_k/\partial a_{ij}$, we simply take $e^T = (0, \ldots, 0, 1, 0, \ldots, 0)$ where 1 occupies the *kth* place. Then $\partial A/\partial a_{ij}$ becomes a matrix the elements of which are all zero except unity in place of a_{ij}. Hence

$$\frac{\partial x_k}{\partial a_{ij}} = y_i x_j$$

which is exactly what Director and Rohrer (1969a, 1969b) invariably did in analyzing sensitivity of linear systems. To summarize their method, we can say that they represented in two boxes the original network N and its adjoint $\hat{N}$, viz.

$Ai_b = 0$ $Bv_b = 0$	$AB^T = 0$ $v_b = Zi_b$	$A\hat{i}_b = 0$ $B\hat{v}_b = 0$
N		$\hat{N}$

A and B are respectively the cut-set and the tie-set matrices, each of which being composed of linearly independent rows, and each satisfying — from graph theory — the identity $AB^T = 0$. i_b and v_b are respectively the branch currents and branch voltages, including as well branch sources. Z is the impedances' matrix. It is diagonal in most cases unless controlled sources are present.

Now, from the relation $Bv_b = 0$, we have that $v_b = A^T V$; V being the nodal voltage vector. Therefore,

$$\langle v_b, i_b \rangle = V^T A i_b = 0$$

Note that here, not only is $\langle v_b, i_b \rangle$ null, but also, and more generally,

$$\langle v_b, \hat{i}_b \rangle = \langle \hat{v}_b, i_b \rangle = 0$$

because both N and $\hat{N}$ have the same configuration, or Topology; i.e. the same matrices A and B. This rule is referred to in network theory as *Tellegen's Theorem*.

Consider now the three circuits N, $\hat{N}$ and N_Δ. The last one, N_Δ, is an incremented circuit, where all variables p_{ij} have been altered to become $p_{ij} + \Delta p_{ij}$.[1] All three circuits satisfy Tellegen's Theorem, since all have the same Topology, same node and branch numbering, same reference directions, etc . . . They only differ in the values of their generators, which will be determined as analysis requires. When we apply Tellegen's theorem to N, $\hat{N}$ and N_Δ, we get, as stated above,

$$\langle v_b, \hat{i}_b \rangle - \langle \hat{v}_b, i_b \rangle = 0$$

because both quantities are null. For N_Δ, we have also that

$$\langle v_b + \Delta v_b, \hat{i}_b \rangle = 0$$

$$\langle \hat{v}_b, i_b + \Delta i_b \rangle = 0$$

i.e.

$$\langle v_b + \Delta v_b, \hat{i}_b \rangle - \langle \hat{v}_b, i_b + \Delta i_b \rangle = 0$$

Upon subtracting the two results, we get

$$\langle \Delta v_b, \hat{i}_b \rangle - \langle \hat{v}_b, \Delta i_b \rangle = 0$$

Now, writing the above relation for all branches while isolating the sources, we get:

$$\underbrace{(\Delta e^T \hat{i}_e - \hat{e}^T \Delta i_e)}_{\text{voltage source}} + (\Delta v_b^T \hat{i}_b - \hat{v}_b^T \Delta i_b) + \underbrace{(\Delta v_j^T \hat{j} - \hat{v}_j^T \Delta j)}_{\text{current source}} = 0$$

where e and j stand for voltage and current sources respectively. But since we assume constant sources, then:

$$\Delta e = 0 \,, \qquad \Delta j = 0$$

Also, from the relation $v_b = Z i_b$, we have that:

$$\Delta v_b = \Delta Z i_b + Z \, \Delta i_b$$

---

1 Writting p_{ij} instead of only p_i has the advantage of allowing for controlled sources. Special cases with Z diagonal exist, like in memory-less circuits where $z_{ii} = p_i$, p_i being the value of the *ith* resistor; or in reactive circuits where $z_{ii} = j\omega p_i$, p_i being in this case the value of the *ith* reactive element. Subsequently, we will take $z_{ij} = p_{ij}$ without loss of generality. Therefore, differentiating the matrix Z with respect to p_{ij} is equivalent to its differentiation with respect to z_{ij}. In the frequency domain, we add only the factor $j\omega$

Here, we have neglected the term $\Delta Z\,\Delta i_b$ since it will be cancelled out anyway in the final derivation. Hence, we obtain;

$$\begin{aligned} -\hat{e}^T \Delta i_e + \Delta v_j^T \hat{j} &= -(\Delta Z i_b + Z\,\Delta i_b)^T \hat{i}_b + \hat{v}_b^T \Delta i_b \\ &= -\hat{i}_b^T(\Delta Z i_b + Z\,\Delta i_b) + \hat{v}_b^T \Delta i_b \\ &= -\hat{i}_b^T \Delta Z i_b + (\hat{v}_b^T - \hat{i}_b^T Z)\,\Delta i_b \end{aligned}$$

Note that in the frequency domain, the transpose is replaced by the conjugate transpose. The last relation obtained is the final result, since by setting $\hat{v}_b^T - \hat{i}_b^T Z = 0$, we define the adjoint network, i.e. a network the impedance matrix of which is transposed. If we focus our interest on $\partial v_j^T/\partial p_{ij}$ — the sensitivity of the output voltage with respect to the variable p_{ij}, we set $\hat{e}_l = 0$, and all $\hat{j}_l = 0$, except for $\hat{j}_k$ set equal to 1. On the other hand, had we been interested in $\partial i_e/\partial p_{ij}$, we would have had to set $\hat{j}_l = 0$, and all $\hat{e}_l = 0$, except for $\hat{e}_k$ set equal to -1. Therefrom follows a simple result, namely that

$$\frac{\partial v_{jk}(\text{or } \partial i_{ek})}{\partial p_{ij}} = -\hat{i}_b^T \frac{\partial Z}{\partial p_{ij}} i_b = -\hat{i}_{bi} i_{bj}$$

In the frequency domain, this result is just multiplied by $j\omega$ ($j = \sqrt{-1}$).

By now, the reader will have noticed the great similarity, or rather the almost exact equivalence, between the above result and the one obtained for the set $Ax = b$. Analysis of N and $\hat{N}$ yields all sensitivity relations between i_{ek} and the elements p_{ij}. Likewise for $Ax = b$, one solution for x and another for y — which belongs to the adjoint system $A^T y = -e$ (e defined a-priori for a particular x_k) — yield the sensitivity relations of x_k with respect to all elements a_{ij}. Note also the resemblance between the vector e defined for $Ax = b$ and $\hat{e}$ used in analysing $\hat{N}$. All elements of e and $\hat{e}$ are null, except for the one element corresponding to the variable Δx_k (in $Ax = b$) and Δi_{ek} (in N).

As a conclusion for the section, we will borrow a simple example from Temes and LaPatra (1977) to clarify the above results. The following figures depict the systems N and $\hat{N}$

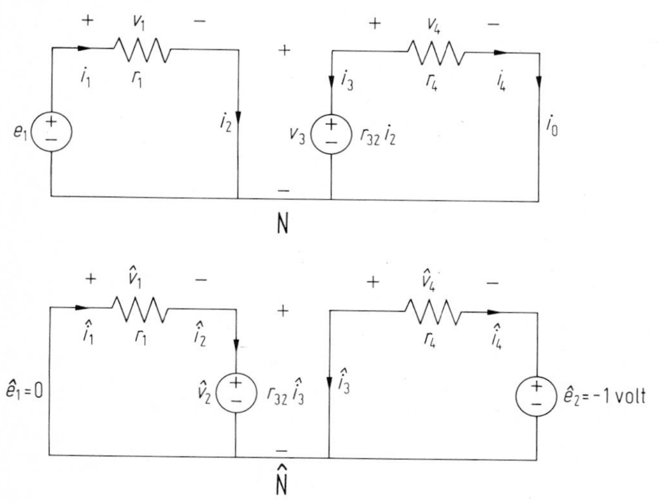

whence

$$Z = \begin{bmatrix} r_1 & 0 & 0 & 0 \\ 0 & 0 & 0 & 0 \\ 0 & r_{32} & 0 & 0 \\ 0 & 0 & 0 & r_4 \end{bmatrix}, \quad \hat{Z} = \begin{bmatrix} r_1 & 0 & 0 & 0 \\ 0 & 0 & r_{32} & 0 \\ 0 & 0 & 0 & 0 \\ 0 & 0 & 0 & r_4 \end{bmatrix}$$

Z is not diagonal because of the existence of a current-controlled voltage source. Now:

$$i_1 = i_2 = e_1/r_1$$

and

$$i_3 = -i_4 = -r_{32}i_2/r_4$$

In the network $\hat{N}$, setting

$$\hat{e}_1 = 0\,, \qquad \hat{e}_2 = -1 \text{ volt}$$

and

$$\hat{i}_3 = -\hat{i}_4 = \hat{e}_2/r_4 = -1/r_4$$
$$\hat{i}_1 = \hat{i}_2 = -r_{32}\hat{i}_3/r_1$$

the sensitivity relations become

$$\frac{\partial i_0}{\partial r_1} = -\hat{i}_1 i_1 = -\frac{e_1 r_{32}}{r_1^2 r_4}$$

$$\frac{\partial i_0}{\partial r_4} = -\hat{i}_4 i_4 = -\frac{e_1 r_{32}}{r_1 r_4^2}$$

$$\frac{\partial i_0}{\partial r_{32}} = -\hat{i}_3 i_2 = \frac{e_1}{r_1 r_4}$$

For further reading on sensitivity analysis of linear systems, the reader is referred to the explanatory work by Calahan (1972).

1.7 An Application on Error Bounds: Lyapunov's Equation

The Lyapunov equation, and more generally the Sylvester equation

$$AX - XB = C$$

appears frequently in the theory of linear dynamic systems, in relation to problems of control and stability. It is solvable using algorithms available for the solution of linear equations, since it can be written

$$(A \otimes I - I \otimes B^T)\, x = c$$

where x and c are respectively vectors containing all the elements of X and C, set in a suitable arrangement. This form defines the *Kronecker product* of two matrices, noted $\otimes$, (see Bellman (1970, Chap. 12)). Hence, solving the Lyapunov equation for X is in fact solving it for x. Only, when A and B are of order m and n, the number of linear equations rises to $m \times n$. For matrices A and B of high order, the algorithms become cumbersome and time-consuming. Bartels and Stewart (1972) set an equivalence between the Sylvester equation and the problem

$$T^{-1}ATT^{-1}XS - T^{-1}XSS^{-1}BS = T^{-1}CS$$

where T and S transform A and B into Jordan form (Schür form for unitary T and S). $T^{-1}XS$ is therefore easily found, wherefrom X could be evaluated. Further formulations of the problem can also be found in Golub, Nash and Van Loan's work (1979).

In this section, the sensitivity of X to changes in A, B and C will be our main concern. At first, we will assume that the unperturbed problem has a unique solution X, a fact guaranteed if

$$\lambda_i(A) - \lambda_j(B) \neq 0\,, \qquad \forall i, j$$

where λ stands for eigenvalue. If A, B and C are subject to perturbations ΔA, ΔB and ΔC, then X will change to $X + \Delta X$. Therefore:

$$A\,\Delta X - \Delta X B = \Delta C - \Delta A X + X\,\Delta B - \Delta A\,\Delta X + \Delta X\,\Delta B$$

i.e.

$$\|A\,\Delta X - \Delta X B\| \leqq \|\Delta C\| + \|X\|\,(\|\Delta A\| + \|\Delta B\|) + \|\Delta X\|\,(\|\Delta A\| + \|\Delta B\|)$$

But we have that

$$\|A\,\Delta X - \Delta X B\| \geq \frac{\min\limits_{i,j} |\lambda_i(A) - \lambda_j(B)|}{\text{cond}\,(T)\ \text{cond}\,(S)}\,\|\Delta X\|$$

where T and S are similarity transformations which diagonalize A and B. For defective A and B, the above bound becomes more complicated. The derivation of the above bound follows directly from Bartels and Stewart's equivalence problem. And by using the relation

$$(\|A\| + \|B\|)\,\|X\| \geqq \|C\|$$

we finally reach

$$\frac{\|\Delta X\|}{\|X\|} \leq \frac{\left\{(\|A\| + \|B\|)\dfrac{\text{cond}\,(T)\ \text{cond}\,(S)}{\min\limits_{i,j} |\lambda_i(A) - \lambda_j(B)|}\right\}\left\{\dfrac{\|\Delta C\|}{\|C\|} + \dfrac{\|\Delta A\| + \|\Delta B\|}{\|A\| + \|B\|}\right\}}{1 - \left\{(\|A\| + \|B\|)\dfrac{\text{cond}\,(T)\ \text{cond}\,(S)}{\min\limits_{i,j} |\lambda_i(A) - \lambda_j(B)|}\right\}\left\{\dfrac{\|\Delta A\| + \|\Delta B\|}{\|A\| + \|B\|}\right\}}$$

provided that the bracket in the denominator is less than unity. Note the similarity between this bound and Wilkinson's (c.f. Sect. 1.4). Here the quantity

$$(\|A\| + \|B\|) \frac{\text{cond}(T)\,\text{cond}(S)}{\min\limits_{i,j} |\lambda_i(A) - \lambda_j(B)|}$$

stands for some condition number (cf. Deif (1983a)). A more formal representation for the condition number is

$$\text{cond}(A, B) = \sup_{\|X\| = \|Y\| = 1} \frac{\|AX - XB\|}{\|AY - YB\|}$$

which is lesser in value than the foregoing expression, since

$$\sup_{\|X\| = 1} \|AX - XB\| \leq \|A\| + \|B\|$$

and

$$\operatorname*{Inf}_{\|Y\| = 1} \|AY - YB\| \geq \min_{i,j} |\lambda_i(A) - \lambda_j(B)| / \text{cond}(T)\,\text{cond}(S)$$

there being no possible equality for any X and Y. The latter infimum is termed, according to Stewart (1971), the *separation* between the two matrices A and B. A special case occurs when both A and B are normal, giving

$$\sup_{\|X\|_2 = \|Y\|_2 = 1} \frac{\|AX - XB\|_2}{\|AY - YB\|_2} = \frac{\max\limits_{i,j} |\lambda_i(A) - \lambda_j(B)|}{\min\limits_{i,j} |\lambda_i(A) - \lambda_j(B)|}$$

since $\text{cond}_2(T) = \text{cond}_2(S) = 1$, T and S being both unitary. For arbitrary matrices A and B, however, the condition number is given from the singular values of $A \otimes I - I \otimes B^T$ as

$$\text{cond}_2(A, B) = \frac{\max\limits_i \sigma_i(A \otimes I - I \otimes B^T)}{\min\limits_i \sigma_i(A \otimes I - I \otimes B^T)}$$

a form in general expensive to calculate. As for the simple case when both A and B are normal, one has in l_2-norm

$$\frac{\|\Delta X\|}{\|X + \Delta X\|} \leq \frac{\|\Delta A\| + \|\Delta B\|}{\min\limits_{i,j} |\lambda_i(A) - \lambda_j(B)|} \quad (\Delta C = 0),$$

a result which coincides with one obtained independently by Jonckheere (1984). Note that in the above bound, the deviations ΔA and ΔB can be large as was already observed by Forsythe and Moler (1967) concerning the equivalent bound related to the problem $Ax = b$ (see Sect. 1.4 and also exercise 1.19).

Beside exploring sensitivity matters, this chapter has cast a light on the value of cond (A) in determining the influence on the solution of alterations in the given data. As we have seen, cond (A) is a measure of the sensitivity of the solution x (of $Ax = b$) to changes in A; be they induced, or unavoidably generated in the course of computations. We have also found that any formulation of a bound for $Ax = b$ must incorporate the notion of conditioning: a measure of how rounding, truncation or perturbation errors are exaggerated in the results due to disparity in the size of the elements of A. Or as Kahan (1966) rightly put it "If A's condition number cond (A) is very large and if A and b are uncertain by a few units in their last place, then no numerical method is capable of solving $Ax = b$ more accurately than to about cond (A) units in x's last place" (cf. exercise 1.12).

However, rounding and truncation are by no means the sole sources of error, for inaccurate data can generate errors exceeding by far the combined effects of the latter two. In practice, this might be the case in physical models (engineering, econometrics, etc, . . .) where data are collected through inaccurate measurements, either with a certain tolerance, or having a mean and a variance. In this case, the coefficients of A and the elements of b are known within a certain accuracy. Thus, they belong to some intervals of expected values, the properties of which we will next proceed to investigate through a study of interval analysis.

Exercises 1

1. Show that cond $(A) = \sigma_1(A)/\sigma_n(A)$ based on l_2-norm, where σ_1 and σ_n are the maximum and minimum singular values of A. Investigate the case where A is normal.

2. If $A = \begin{bmatrix} 1 & 3 \\ 1 & 4 \end{bmatrix}$, and $A = \begin{bmatrix} 1 & 1 \\ 0.1 & 0.101 \end{bmatrix}$, discuss cond (A) in each case.

3. Show that cond $(AB) \leqq$ cond (A) cond (B).

4. If U is unitary, show that $\text{cond}_2\,(UA) = \text{cond}_2\,(AU) = \text{cond}_2\,(A)$.

5. Comment on $A = \text{diag}\,(10^5, 10^{-5})$, $\det(A) = 1$, cond $(A) = 10^{10}$. Find the nearest singular matrix.

6. If $\|B\|/\|A\| = \theta < 1$, show that cond $(A + B) \leqq (\text{cond}\,(A) + \theta)/(1 - \theta)$.

7. If $\det(A) \to 0$, show that cond $(A) \to \infty$. Hint: consider $\min_v \dfrac{\|Av\|}{\|v\|}$ taking $v_k = C_{ik}$, C_{ik} being the cofactor of a_{ik}.

8. For $A = \begin{bmatrix} \alpha + \beta & \alpha \\ \alpha & \alpha - \beta \end{bmatrix}$, show that det (A) and cond (A) can be made independent. Hint: $\det(A) = -\beta^2$, cond $(A) = 1 + \dfrac{2}{\beta^2}(\alpha^2 + |\alpha|\sqrt{\alpha^2 + \beta^2})$ grows with $|\alpha|$ growing.

9. For $A = \begin{bmatrix} \alpha + \beta & \alpha \\ \alpha & \alpha + \beta \end{bmatrix}$, show that det (A) and cond (A) are interdependent.

   Hint: $\det(A) = \beta(2\alpha + \beta)$, cond $(A) = 1 + 2\left|\dfrac{\alpha}{\beta}\right|$ increasing simultaneously with α growing.

10. Show that $\det(I + \varepsilon A) = \exp\left(\sum_{k=1}^{\infty} \frac{(-)^{k-1}}{k} \varepsilon^k \operatorname{tr} A^k\right)$. Hence obtain for A nonsingular that $\det(A + \varepsilon B) = \det(A) + \varepsilon \operatorname{tr}(A^a B) + O(\varepsilon^2)$.

11. Show that $\det(I + \varepsilon B) \leq \left(1 + \varepsilon \frac{\operatorname{tr} B}{n}\right)^n$, with n = order of B. Are there any restrictions on B? Hint: $\det(A) = \prod_i \lambda_i$ (λ is the eigenvalue), $\operatorname{tr}(A) = \sum_i \lambda_i$, use relation between arithmetic and geometric mean.

12. If $\operatorname{cond}(A) \cong 10^p$, show that the solution $\hat{x}$ of the linear system $Ax = b$ computed in t-digit(decimal)arithmetic has almost t-p significant digits. Check your result by solving the equations

$$1.332x + 0.664y = 1.996\,, \qquad 0.665x + 0.334y = 0.999$$

on a 10-digit machine and having $\operatorname{cond}(A) \cong 10^3$

13. If for the first numerical example in Sect. 1.3 of the 3×3 matrix A, A is required to be nonsingular on a t-digit machine, obtain a lower bound for t.

14. Compute a null vector for $A = \begin{bmatrix} -1 & 4 & -2 \\ 0 & -1 & 3 \\ 2 & -9 & 7 \end{bmatrix}$

15. Show that for A positive definite, B real symmetric

$$(A + \varepsilon B)^{-1} = A^{-1/2}(I + \varepsilon S)^{-1} A^{-1/2}$$

where S is symmetric. Thus obtain the symmetric perturbation formula

$$(A + \varepsilon B)^{-1} = A^{-1} - \varepsilon A^{-1/2} S A^{-1/2} + \ldots$$

16. Solve $\begin{bmatrix} 8 & 1 \\ 1 & 1 \end{bmatrix} \begin{bmatrix} x_1 \\ x_2 \end{bmatrix} = \begin{bmatrix} 1 \\ -1 \end{bmatrix}$

on a three-digit machine. Obtain a bound for $\|\Delta x\|/\|x\|$ in terms of the round-off errors.

17. If $Ax = b$, with $A = \begin{bmatrix} 1 & 2 \\ 3 & 4 \end{bmatrix}$, $b = \begin{bmatrix} 5 \\ 6 \end{bmatrix}$ and if A and b exhibit respectively changes $\Delta A = \begin{bmatrix} 0.1 & 0 \\ 0.02 & 0.3 \end{bmatrix}$, $\Delta b = \begin{bmatrix} 0.01 \\ -0.1 \end{bmatrix}$. Evaluate an approximate bound for $\|\Delta x\|/\|x\|$.

18. Show that if $\|A^{-1}\,\Delta A\| < 1$ then:

$$a - \|(A + \Delta A)^{-1}\| \leq \frac{\|A^{-1}\|}{1 - \|A^{-1}\,\Delta A\|}$$

$$b - \frac{\|(A + \Delta A)^{-1} - A^{-1}\|}{\|A^{-1}\|} \leq \frac{\|A^{-1}\,\Delta A\|}{1 - \|A^{-1}\,\Delta A\|}$$

$$c - \|(A + \Delta A)^{-1}\| \geq \|A^{-1}\| \left(1 - \frac{\|A^{-1}\,\Delta A\|}{1 - \|A^{-1}\,\Delta A\|}\right)$$

19. Using the identity $B^{-1} - A^{-1} = -A^{-1}(B - A)\,B^{-1}$, obtain that

$$\frac{\|(A + \Delta A)^{-1} - A^{-1}\|}{\|(A + \Delta A)^{-1}\|} \leqq \|A^{-1}\,\Delta A\|$$

And by writing $\Delta x = -A^{-1}(B - A)\,B^{-1}b$ if $\Delta b = 0$, show similarly that

$$\frac{\|\Delta x\|}{\|x + \Delta x\|} \leqq \|A^{-1}\,\Delta A\|$$

20. Let B be an approximate inverse of A due to a machine accuracy ε, show that a rule of thumb of the computed B is:
(number of correct decimal digits) $\geqq$ (number of digits carried)
$- \log(\|A\|\;\|A^{-1}\|) - \log(10n)$,
where $n =$ dimension of A.

21. The condition number cond (A) refers to the sensitivity of the problem discussed. Many variations of the condition number exist for the various problems. Show that, for the simple problem of multiplying two matrices A and B, i.e. $C = AB$, the sensitivity of C with respect to the variations in A and B is governed by

$$\frac{\|\Delta C\|}{\|C\|} \leq \frac{\|A\|\;\|B\|}{\|AB\|}\left(\frac{\|\Delta A\|}{\|A\|} + \frac{\|\Delta B\|}{\|B\|}\right)$$

Define the condition number for the problem.

22. For the equations $0.2161x_1 + 0.1441x_2 = 0.144$, $1.2969x_1 + 0.8648x_2 = 0.8642$ having $x_1 = 2$, $x_2 = -2$ exactly, if $\hat{x}_1 = 0.9911$, $\hat{x}_2 = -0.4870$ are approximate solutions with residuals $r_1 = -0.00000001$ and $r_2 = 0.00000001$, explain why r_1 and r_2 are small though $\hat{x}_1$ and $\hat{x}_2$ are by far inaccurate.

23. Consider the equations $x_1 + x_2 = 2$, $x_1 + 1.00001x_2 = 2.00001$ having $x_1 = x_2 = 1$, select $\hat{x}_1 = 1 + \alpha$, $\hat{x}_2 = 1 + \beta$ and obtain $r_1 = -\alpha - \beta$, $r_2 = -\alpha - 1.00001\beta$. Show by taking $\alpha + \beta = 0$ that r_1 and r_2 can be made arbitrary small, while $\hat{x}_1$ and $\hat{x}_2$ becoming very inaccurate.

24. Show that the system

$$2x_1 + x_2 + x_3 = 1$$
$$x_1 + \varepsilon x_2 + \varepsilon x_3 = 2\varepsilon$$
$$x_1 + \varepsilon x_2 - \varepsilon x_3 = \varepsilon$$

having

$$\det(A) = 2\varepsilon(1-2\varepsilon), \qquad A^{-1} = \begin{bmatrix} \frac{-\varepsilon}{1-2\varepsilon} & \frac{1}{1-2\varepsilon} & 0 \\ \frac{1}{1-2\varepsilon} & \frac{-1-2\varepsilon}{2\varepsilon(1-2\varepsilon)} & \frac{1}{2\varepsilon} \\ 0 & \frac{1}{2\varepsilon} & \frac{-1}{2\varepsilon} \end{bmatrix}$$

and $x_1 = \dfrac{\varepsilon}{1-2\varepsilon}$, $x_2 = \dfrac{1}{2} - \dfrac{2\varepsilon}{1-2\varepsilon}$, $x_3 = \dfrac{1}{2}$, is well-conditioned

25. The following linear system is obtained for a bridged T network

$$\begin{bmatrix} \frac{1}{R_s} + \frac{1}{R_1} + \frac{1}{R_4} & -\frac{1}{R_1} & -\frac{1}{R_4} \\ -\frac{1}{R_1} & \frac{1}{R_1} + \frac{1}{R_2} + \frac{1}{R_3} & -\frac{1}{R_2} \\ -\frac{1}{R_4} & -\frac{1}{R_2} & \frac{1}{R_0} + \frac{1}{R_4} + \frac{1}{R_2} \end{bmatrix} \begin{bmatrix} v_1 \\ v_2 \\ v_3 \end{bmatrix} = \begin{bmatrix} I \\ 0 \\ 0 \end{bmatrix}$$

obtain $\partial v_3/\partial R_3$ at $R_s = R_3 = R_0 = R_4 = 1\ \text{K}\Omega$, $R_1 = R_2 = 0.5\ \text{K}\Omega$ and $I = 1$ mA. Find also ΔR_4 which compensates for an error in R_3 given by $\Delta R_3 = 1\ \Omega$ using first order approximation so that $\Delta v_3 = 0$.

26. Show that the results in Sect. 1.6 concerning sensitivity analysis of a linear system can be modified to account for all kinds of controlled sources. Hint: carry out the same analysis using a "Hybrid branch matrix" H made from the relations:

$$(i_{b1} \mid v_{b2})^T = H(v_{b1} \mid i_{b2})^T$$

27. For the circuit shown draw $\hat{N}$ and hence derive an expression for

$$\frac{\partial v_0}{\partial L, \partial C, \partial R}$$

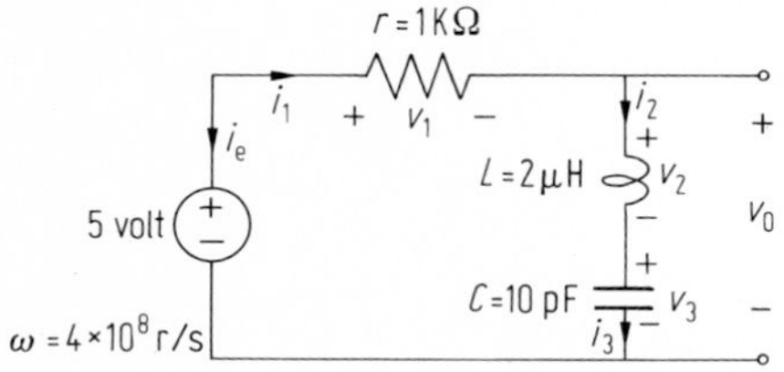

r = 1KΩ
i_1 + V_1 −
i_e
i_2 + V_2 −
L = 2μH
5 volt
$\omega = 4 \times 10^8$ r/s
C = 10 pF
+ V_3 −
i_3
+ V_0 −

28. Obtain a bound for $\|\Delta X\|/\|X\|$, if $AX - XB = C$ undergoes a perturbation with:

$$A = 1, \quad B = \begin{bmatrix} 6 & 8 \\ 8 & -6 \end{bmatrix}, \quad C = [0 \quad 1]$$

$$\Delta A = 0, \quad \Delta B = \begin{bmatrix} 0.1 & -0.2 \\ 0.3 & 0 \end{bmatrix}, \quad \Delta C = [0.1 \quad 0]$$

29. Show that

$$\operatorname{cond}(A \otimes I + I \otimes B^T) = \alpha \frac{\max_{i,j} |\lambda_i(A) + \lambda_j(B)|}{\min_{i,j} |\lambda_i(A) + \lambda_j(B)|}$$

where α satisfies

$$1/\operatorname{cond}^2(T)\operatorname{cond}^2(S) \leqq \alpha \leqq \operatorname{cond}^2(T)\operatorname{cond}^2(S)$$

and where T and S diagonalize A and B.

30. Consider $X = C + \varepsilon(AX + XB)$ and by writing $X = C + \sum_{n=1}^{\infty} \varepsilon^n \Phi_n(A, B)$, show that $\varphi_n = A\varphi_{n-1} + \varphi_{n-1}B$, $\varphi_0 = C$, and that $\Phi_n = A^n C + \binom{n}{1} A^{n-1}CB + \ldots + CB^n$. Moreover X can be written as $X = P([I - \varepsilon(A + B)]^{-1} C)$ with P suitably chosen.

Chapter 2

Methods of Interval Analysis

2.1 Introduction

Three types of errors are encountered in numerical analysis, namely:

1. Round-off errors, arising when numbers are rounded to fit a certain precision arithmetic; e.g. the case where $1/6 = 0.1666 \ldots$ is approximated to 0.167 on a three-digit machine.
2. Truncation errors, resulting when convergent series are truncated down to a number of terms, e.g. the case where $\pi = 3.14159265 \ldots$ is approximated by $\pi = 3.14$.
3. Data errors, associated with the specific physical model under study. They represent a parameter's uncertainties when it is determined through experimental measurements.

The first two types can be remedied almost completely either by increasing the machine's precision in the first case, or by indulging in lengthy computations for the second. Data errors, for their part, are uncontrollable. In solving, for instance, the differential equation

$$\frac{dy}{dx} = ax^b, \qquad x(0) = c$$

where the constants a, b and c include a certain uncertainty, we would get a solution y which is itself uncertain. Interval analysis deals incidentally with evaluating the error in y resulting from errors in the constants a, b and c.

A problem to be investigated in this chapter is formulated hereafter. In solving the equations $Ax = b$, if each element a_{ij} of A is allowed to vary around a mean or centre value (denoted a_{ij}^c or $m(a_{ij})$) within certain bounds, that is

$$a_{ij}^c - \varepsilon_{ij} \leqq a_{ij} \leqq a_{ij}^c + \varepsilon_{ij}$$

then what would be the expected variations in x? Here, ε_{ij} is the maximum error (or uncertainty) anticipated in the element a_{ij} of the data. Finding these upper and lower bounds for x is not such an easy matter, for they can surely not be evaluated through the standard error analysis discussed in chapter one. In fact, finding x using the formula:

$$x + \Delta x = (A + \Delta A)^{-1}\, b$$

would necessitate the expansion of $(A + \Delta A)^{-1}$ into an infinite series in $\Delta A(\varrho(A^{-1}\,\Delta A) < 1)$. Substituting by $\Delta A = \pm E$ (an error matrix) would yield $x \pm \Delta x$ only by considering all terms. To emphasize the futility of this procedure, consider for instance the solution of the overly simplistic problem

$$ax = b\,, \qquad a = \frac{3}{2} \pm \frac{1}{2} \quad \text{and} \quad b = 1$$

Then $x = x^c + \dfrac{dx}{da}\Delta a + \dfrac{d^2x}{da^2}\dfrac{(\Delta a)^2}{2!} + \ldots$

with

$$\frac{\mathrm{d}x}{\mathrm{d}a} = -\frac{b}{a^2}\,, \qquad \frac{\mathrm{d}^2x}{\mathrm{d}a^2} = \frac{2b}{a^3}\,, \ldots$$

whence

$$x = \frac{2}{3} - \frac{1}{(3/2)^2}\left(\pm\frac{1}{2}\right) + \frac{(2)\,(1)}{(3/2)^3}\,\frac{\left(\pm\frac{1}{2}\right)^2}{2!} + \ldots$$

$$= \frac{2}{3} \pm \frac{2}{9} + \frac{2}{27} + \ldots \cong \frac{20}{27} \pm \frac{2}{9}$$

i.e. x is included approximately between the bounds

$$\frac{14}{27} < x < \frac{26}{27}$$

On the other hand, writing the equation directly as

$$[1, 2]\, x = 1$$

i.e.

$$a \in [1, 2]$$

yields directly

$$x \in [1/2, 1/1] = [1/2, 1]$$

which are the exact bounds.

Standard error analysis is usually used when the perturbation ΔA is finite and fixed in value, so that Δx is also finite and uniquely defined. It may also be used when $|\Delta a_{ij}|$ is small enough, so that only the 1st order perturbations are considered. In case the data lie within a certain range of values, an adequate method should be designed to analyse directly the exact ranges instead of just the parameter variations or perturbations. This approach is referred to as *interval analysis*. It yields a solution in the form of bounds similar to those obtained in the previous example. It is not compulsory however that data errors be given in the form of bounds, but they should be expressed in this form whenever possible. Rounding and truncation errors, for instance, can be dealt with through interval analysis, since the number $1/6 = 0.1666\ldots$ can be written as

$$1/6 \in [0.166, 0.167]$$

and likewise

$$\pi = 3.1415\ldots \in [3.14, 3.15]$$

Interval analysis can therefore substitute for standard error analysis. It suffers however from one major drawback, namely that its arithmetics do not conform with the common rules of real arithmetic. This yields larger-than-expected error bounds, when numerical analysts' concern lies more in finding *bound-conserving* algorithms than in finding stable ones as in standard error analysis. Furthermore, apart from yielding pessimistic bounds, interval analysis is time consuming. To the first criticism, workers in the field retort that it is remedied by extensive computation; to the second, they respond by suggesting that computer hardware be preprogrammed with interval arithmetic so that its computations can be executed at speeds comparable to those of ordinary machine arithmetic. A breakthrough was apparently made in this area by Kulisch and his coworkers (see Kulisch and Miranker (1983)). In any case, interval analysis stands now as a new branch in mathematics. Since the work performed early on by Moore, Hansen and their coworkers in the U.S.A., in the early sixties, and by Krückeberg, Nickel and their colleagues in Germany, interval analysis has come a long way, and at a furious pace. Nowadays, one would hear of such topics as interval topology, interval calculus, interval geometry, etc . . ., together with other applications ranging from electrical circuits to psychology. Moore's second book (1979) contains a large bibliography, together with a large appendix — edited by Bierbaum and Schwiertz — containing an exhaustive listing of the work published uptill 1978, thus updating Bierbaum (1974, 1975). An updated version of the interval analysis library is also compiled by Garloff (1985). Early introductions to this topic were Moore's first book (1966), Alefeld and Herzberger's book (1974) in German, with its subsequent English edition (1983). We can also quote the proceedings of two conferences, held in Oxford in 1968 and in Karlsruhe in 1975, edited respectively by Hansen (1969) and Nickel (1975). A third conference also took place in Freiburg in 1980 and its proceedings was edited by Nickel (1980).

2.2 Interval Arithmetic

Setting $x \in [a, b]$ signifies that x is allowed to take any value between a and b, including both end points. It also implies that $b \geqq a$. For this reason, some authors opt for the notation $[\underline{a}, \overline{a}]$ for an interval $[a]$ of which the lower and upper bounds are respectively $\underline{a}$ and $\overline{a}$. When $\underline{a} = \overline{a}$, $[a]$ becomes an interval with zero segment, i.e. a point in $\mathbb{R}^1$. It is then referred to as a *point or degenerate interval*; it can alternatively be noted a. Hence, interval analysis embraces ordinary real point analysis as a special case. When stating that

$$[a] \subseteq [b]$$

we imply that the segment or interval $[\underline{a}, \bar{a}]$ is entirely contained in $[\underline{b}, \bar{b}]$; i.e. $\underline{a} \geqq \underline{b}$ and $\bar{a} \leqq \bar{b}$. On the other hand, $[a] < [b]$ means that $\bar{a} < \underline{b}$. Two intervals $[a]$ and $[b]$ are said to be equal if their end points coincide, i.e. $\underline{a} = \underline{b}$ and $\bar{a} = \bar{b}$. When $[a]$ and $[b]$ have a certain range of values in common, then they are said to intersect; i.e. $[a] \cap [b] \neq \emptyset$. On the other hand, when $\underline{a} > \bar{b}$ or $\bar{a} < \underline{b}$, their intersection is void, and $[a] \cap [b] = \emptyset$.

The difference between $\underline{a}$ and $\bar{a}$ is defined as the *width* or *span* of $[a]$, i.e. $w[a] = \bar{a} - \underline{a}$. This quantity may be taken as a measure of the uncertainty in the variable a. For an interval matrix A^I, the elements of which are $[a_{ij}]$, we take

$$w(A^I) = \max_{i,j} w[a_{ij}]$$

From this short introduction we can now proceed to state some basic rules, namely

$$[a] + [b] = [\underline{a}, \bar{a}] + [\underline{b}, \bar{b}] = [\underline{a} + \underline{b}, \bar{a} + \bar{b}]$$
$$[a] - [b] = [\underline{a}, \bar{a}] - [\underline{b}, \bar{b}] = [\underline{a} - \bar{b}, \bar{a} - \underline{b}]$$
$$[a] \cdot [b] = [\underline{a}, \bar{a}] \cdot [\underline{b}, \bar{b}] = [\min(\underline{a}\underline{b}, \underline{a}\bar{b}, \bar{a}\underline{b}, \bar{a}\bar{b}), \max(\underline{a}\underline{b}, \underline{a}\bar{b}, \bar{a}\underline{b}, \bar{a}\bar{b})]$$
$$[a]/[b] = [\underline{a}, \bar{a}] \cdot \left[\frac{1}{\bar{b}}, \frac{1}{\underline{b}}\right] \quad \text{iff} \quad 0 \notin [\underline{b}, \bar{b}]$$

As it turns out to be, interval addition and multiplication are associative and commutative. As for the distributivity, consider the instance

$$[1, 2]([1, 2] - [1, 2]) \neq [1, 2][1, 2] - [1, 2][1, 2]$$

where the left-hand side equals $[-2, 2]$, and the right-hand side equals $[-3, 3]$. However $[-2, 2] \subset [-3, 3]$, which expresses a property known as *subdistributivity*; stated as

$$[a]([b] + [c]) \subseteq [a][b] + [a][c]$$

for the three intervals $[a]$, $[b]$ and $[c]$ (cf. Moore (1966)). Fortunately, the cancellation law holds, for if $[a] + [b] = [a] + [c]$ then $[b] = [c]$. Then again, having that

$$[a] + [x] = [b]$$

does not entail that $[x] = [b] - [a]$. For instance, we have that

$$[1, 2] + [3, 4] = [4, 6]$$

and

$$[4, 6] - [1, 2] = [2, 5]$$

For a review of the rules of interval algebra and logic, the reader is referred to Ratschek (1975).

This oddness of the rules complicates matters seriously. Indeed it is very disappointing to note that for one and the same interval $[a]$

$$[a] - [a] \neq 0$$
$$[a] \ / \ [a] \neq 1$$
$$[a] \cdot [a] \neq [a^2] \qquad \text{(when } 0 \in [a])$$

It requires much effort of an analyst to devise techniques fit to deal with so many odd rules — so many obstacles. And, unless interval computations are carried out with utmost care, results could turn out to be erroneous and misleading.

To demonstrate the above rules of interval arithmetic, let us borrow a simple example from Moore (1969). We wish to evaluate

$$y = \frac{a_1 + a_2 x}{a_3 + a_4 x^2}$$

when $a_1 = 0.2 \pm 0.001$; $a_2 = 0.3 \pm 0.005$; $a_3 = 6.17 \pm 0.02$; $a_4 = -2 \pm 0.1$ and $x = 0.452 \pm 0.001$. When performed step by step on a three-decimal-digit machine, the calculations are as follows:

$$[a_1] = [0.199, 0.201]$$
$$[a_2] = [0.295, 0.305]$$
$$[a_3] = [6.15, 6.19]$$
$$[a_4] = [-2.1, -1.9]$$
$$[x] = [0.451, 0.453]$$
$$[x^2] = [0.203, 0.206]$$
$$[a_4][x^2] = [-0.433, -0.385]$$
$$[a_3] + [a_4][x^2] = [5.71, 5.81]$$
$$[a_2][x] = [0.133, 0.139]$$
$$[a_1] + [a_2][x] = [0.332, 0.340]$$

so that

$$[y] = \frac{[0.332, 0.340]}{[5.71, 5.81]} = [0.0574, 0.0599]$$

i.e.

$$y = 0.0586 \pm 0.0013$$

The above calculations were executed using *rounded-interval arithmetic* in order to account for any eventual round-off errors in the intervals' end points. In fact, although $[a]$ was represented as an interval to account for possible fluctuations around its mean value equal to $\pm\frac{1}{2}w([a])$, the end points of this interval may themselves be subject to some error. This would perpetuate our error in a vicious sequence, a definition of $[a]$ becoming thence impossible to reach. For instance, if $[a] = [3, 6]$, then $1/[a] = [0.166\ldots, 0.333\ldots]$, having inaccurately defined end points. Had we replaced the equal sign by the $\subseteq$ sign, as in rounded-interval arithmetic, we could have carried out the computation as $1/[a] \subseteq [0.166, 0.334]$. A numerical example on the use of rounded-interval arithmetic is also discussed in Moore (1979, p. 16).

In the above example, interval manipulations proved efficient by giving narrow bounds for the solution y. In many cases, however, they may yield loose bounds for the result around its expected value; one famous example for such a situation is Gauss' interval elimination. Suppose $Ax = b$ is to be solved using an appropriate pivoting technique together with interval arithmetic to account for rounding errors unavoidably generated in the course of computation. Let x be the exact solution for the equation, and $x(l)$ the actual solution obtained (here l = word length), then $[x(l)]$ will stand for the corresponding interval approximation. Obviously, $x \in [x(l)]$; the question arises as to how large $w[x(l)]$ is, i.e. how loose the error bounds are. Wilkinson's a priori analysis has shown that

$$\|x - x(l)\| \approx \varepsilon(l) \operatorname{cond}(A)$$

where $\varepsilon(l)$ is a measure of the relative error in the elements of A (cf. Dahl, Sect. 1.5; Wilkinson (1965) p. 197). Furthermore, are the computed bounds in $[x(l)]$ equally favourable? Wongwises (1975) conducted several thousand experiments on various computers. The matrix A was created at random, in order to vary cond (A). The quantity $w[x(l)]/\|x - x(l)\|$ was plotted versus n, the order of A. Wongwises found that this index of uncertainty in the result increases very rapidly with n; for $n = 30$, $w[x(l)]/\|x - x(l)\|$ yields an overestimation of 10^9, which is obviously unacceptable. What is more awkward is that results have nothing to do with cond (A) or the value of l, a conclusion drawn when we are lucky enough to see the algorithm reach its end. More usually than not, computation is interrupted, simply because of a pivot containing the zero in its interval, which is a very probable occurrence.

We will borrow from Hansen (1969) an example in which such large bounds can be obtained. Here, we wish to solve the equations

$$\begin{aligned} [2, 3]\, x_1 + [0, 1]\, x_2 &= [0, 120] \\ [1, 2]\, x_1 + [2, 3]\, x_2 &= [60, 240] \end{aligned}$$

using standard interval arithmetic

$$[x_1] = \frac{[0, 120]\,[2, 3] - [60, 240]\,[0, 1]}{[2, 3]\,[2, 3] - [1, 2]\,[0, 1]} = [-120, 180]$$

$$[x_2] = \frac{[2, 3]\,[60, 240] - [0, 120]\,[1, 2]}{[2, 3]\,[2, 3] - [1, 2]\,[0, 1]} = [-60, 360]$$

We will see shortly why these bounds are loose and misleading. Some improvements can be achieved by rearranging the above quotients as Hansen and Smith (1967) noted. The estimate obtained is still not that much better, notwithstanding the size of the problem. In fact, in the simpler instances, as when we compute the range of values of $a/(a-2)$, $a \in [10, 12]$ we obtain $[1, 3/2]$ when using interval arithmetic, whilst the exact range — $[1.2, 5/4]$ — could be obtained upon inspection, taking note of the quotient's monotonicity. This drawback has been known for a long time to numerical analysts, who realized the inconvenience of solving linear equations using interval arithmetic. Yet, for reasons of bad publicity, as Nickel pointed out (1977) this inconvenience was not admitted widely until Wongwises stipulated it in 1975. Many a numerical analyst remained skeptical, or even suspicious, about this new field for a long time. Nowadays, thanks to innovation and testing of numerous algorithms and the associated progress, interval arithmetic has become an established discipline. Gauss' methods themselves have undergone further progress that narrowed greatly the error bounds.

The main questions are yet to be answered: what are the exact ranges of values for x_1 and x_2? What is the significance of finding a solution x for $A^I x = b^I$, $A^I = [\underline{A}, \overline{A}]$ and $b^I = [\underline{b}, \overline{b}]$ being respectively an interval matrix and an interval vector?

When endeavouring to solve $A^I x = b^I$, we set out to find all the possible values of the vector $x \in \mathbb{R}^n$ satisfying the equation $Ax = b$, where A and b are fixed and assume all possible combinations of values inside A^I and b^I. This infinite number of solutions constitutes a region inside $\mathbb{R}^n$ which we will call X. In other terms, solving $A^I x = b^I$ for x is synonymous to finding X, given by

$$X = \{x : Ax = b\,, \qquad A \in A^I\,, \qquad b \in b^I\}$$

As for the interval vector x^I, it is in fact the vector with the minimum possible interval containing X. Note that we do not usually write $A^I x^I = b^I$, because, as a matter of fact, $A^I x^I \supseteq b^I$. In the sequel, we shall assume that every matrix A contained in A^I is nonsingular.

Now, for $Ax = b$ when A and b take respectively all values of A^I and b^I, the set X containing all possible solutions x is represented by the two inequalities

$$\underline{A}x \leqq \overline{b}$$
$$\overline{A}x \geqq \underline{b}$$

valid for nonnegative values of the elements x_i. A proof of this assertion is provided in a later section. For the simple numerical example given above, the foregoing inequalities will read respectively

$$2x_1 \leqq 120\,, \qquad x_1 + 2x_2 \leqq 240$$
$$3x_1 + x_2 \geqq 0\,, \qquad 2x_1 + 3x_2 \geqq 60$$

Similar inequalities can be obtained for negative values of x_1 and x_2. All these inequalities define a region X in $\mathbb{R}^2$ containing all possible solutions x to the problem. This region we will call the domain of compatible solutions. Obviously, it has the shape of a polygon (see Fig. 1) and although X here is convex in any one quadrant, the union of all polygons lying in all possible quadrants is usually a nonconvex domain. This is the exact approach to the problem, for x^I can easily be shown to be equal to

$$x^I = ([-120,90], [-60,240])^T$$

Note the accuracy of this bound when compared to that obtained using standard interval arithmetic. Again, it is in general difficult to find X, or to represent it once it is obtained; especially in spaces with more dimensions than two. One must thus adopt a different approach.

Solution of linear equations with interval coefficients has followed two main channels, namely:

a) To try and estimate $(A^I)^{-1}$; an approach adopted by Hansen et al. The solution x^I, hopefully the narrowest interval containing X, satisfies the relation $x^I \subseteq (A^I)^{-1}\, b^I$.
b) To use linear programming to define X as

$$X = \{x\colon Ax = b\,, \qquad A \in A^I\,, \qquad b \in b^I\}$$

x satisfies the relation $A^I x \cap b^I \neq \emptyset$. This approach was adopted by Oettli et al. It was further found useful in studying the compatibility of solutions to $Ax = b$, A and b having fixed values.

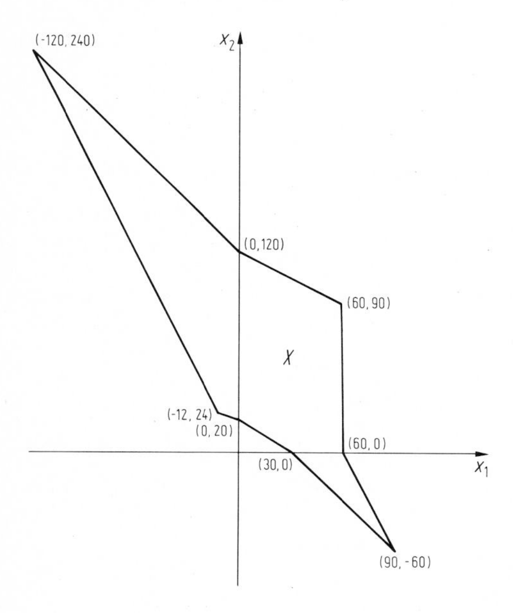

Fig. 2.1

2.3 Hansen's Methods

In their two papers (Hansen (1965); Hansen and Smith (1967)), the authors sought an estimate of $(A^I)^{-1}$ that would include in its intervals the coefficients of

$(A^c)^{-1}$, A^c being the matrix of mean values of the intervals of A^I, termed the centre of A^I, i.e.

$$a_{ij}^c = (\underline{a}_{ij} + \bar{a}_{ij})/2$$

Now, let E^I be an error interval matrix given by

$$E^I = I - A^I B$$

where B is an approximate inverse of A^c, computed using some matrix-inversion algorithm with single-precision floating-point machine arithmetic. E^I has smaller valued elements when A^I has a narrower width. Similar to the bound used for a fixed matrix A, one can assign a norm to E^I or A^I, i.e.

$$\|A^I\| = \max_i \sum_{j=1}^{n} \max(|\underline{a}_{ij}|, |\bar{a}_{ij}|)$$

Now from the above residual relation, we could write

$$(A^I)^{-1} = B(I - E^I)^{-1}$$

which is unfortunately impossible to evaluate. For one thing, one must rely in its evaluation on the identity $(A^I)(A^I)^{-1} = I$; which is not necessarily true. Had it been valid, $(I - E^I)^{-1}$ would not be equal to the series

$$I + E^I + (E^I)^2 + \ldots \qquad (\text{with } \|E^I\| < 1)$$

since a proof of such an identity must rely on the distributive law which does not hold.

Instead, one should make use of the relation

$$E^c = I - A^c B$$

but first obtain

$$\begin{aligned} \|(I - E^c)^{-1} - (I + E^c + (E^c)^2 + \ldots + (E^c)^m\| &\leqq \|(E^c)^{m+1}\| \, \|(I - E^c)^{-1}\| \\ &\leqq \frac{\|E^c\|^{m+1}}{1 - \|E^c\|}, \qquad \|E^c\| < 1 \\ &\leqq \frac{\|E^I\|^{m+1}}{1 - \|E^I\|}, \text{ since } \|E^I\| \geqq \|E^c\| \end{aligned}$$

Meanwhile, note that since $A^c \subseteq A^I$ and $E^c \subseteq E^I$, then

$$\begin{aligned} (A^c)^{-1} = B(I - E^c)^{-1} &\subseteq B(S_m^c + D^I) \\ &\subseteq B(S_m^I + D^I) \end{aligned}$$

Here, D^I is a matrix with identical elements, each of which is given by

$$[d] = \left[\frac{-\|E^I\|^{m+1}}{1 - \|E\|^I}, \frac{\|E^I\|^{m+1}}{1 - \|E^I\|}\right]$$

Furthermore

$$S_m^I = I + E^I(I + E^I(I + E^I(I + E^I(\ldots)))) \text{ to } m \text{ sums};$$

the relation is written in this order to make use of the subdistributivity law. To exemplify the foregoing theoretical discussion, let us borrow one example given by Moore (1966), viz.

$$A^I = \begin{bmatrix} [0.999, 1.01] & [-1.00 \times 10^{-3}, 1.00 \times 10^{-3}] \\ [-1.00 \times 10^{-3}, 1.00 \times 10^{-3}] & [0.999, 1.01] \end{bmatrix}$$

whence $A^c = I$ and $B = I$; and

$$E^I = I - A^I B = \begin{bmatrix} [-0.01, 0.001] & [-0.001, 0.001] \\ [-0.001, 0.001] & [-0.01, 0.001] \end{bmatrix}$$

Here, $\|E^I\| = 0.011$ and, for $m = 1$,

$$0.000\,122 < \frac{\|E^I\|^2}{1 - \|E^I\|} < 0.000\,123$$

Taking the upper bound as a binding value for the interval, we obtain

$$D^I = \begin{bmatrix} [-0.000\,123, 0.000\,123] & [-0.000\,123, \ 0.000\,123] \\ [-0.000\,123, 0.000\,123] & [-0.000\,123, \ 0.000\,123] \end{bmatrix}$$

wherefrom we finally get

$$(A^c)^{-1} \subseteq \begin{bmatrix} [0.9898, 1.001\,13] & [-0.001\,13, 0.001\,13] \\ [-0.001\,13, 0.001\,13] & [0.989\,8, 1.001\,13] \end{bmatrix}$$

If the reader would take $m = 2$ and compare the results, he will readily conclude that the greater the value of m, the higher the accuracy attained. Moore (1966) found out that the upper and lower bounds of any component of the interval matrix $\{(A^c)^{-1}/A^c \subseteq A^I\}$ fall short of being sharp by a quantity of the order of the square of the width of A^I. As for the solution $x^I \subseteq (A^I)^{-1} b^I$, it can be calculated by multiplying $(A^I)^{-1}$ by b^I.

The same procedure can also be applied to a definite-element matrix A, when its elements have rational values. For instance

$$A = \begin{bmatrix} 1 & 1/2 & 1/3 \\ 1/2 & 1/3 & 1/4 \\ 1/3 & 1/4 & 1/5 \end{bmatrix} = \begin{bmatrix} 1 & 0.5 & \alpha \\ 0.5 & \alpha & 0.25 \\ \alpha & 0.25 & 0.2 \end{bmatrix}$$

with $\alpha = [0.3333333333, 0.3333333334]$ in double-precision arithmetic. For this reason, Hansen's method is valuable to numerical analysts when solving the simple problem $Ax = b$. It can replace the error analysis approach by Wilkinson (1963).

Another method, proposed by Hansen and Smith in their 1967 paper and yielding good results for narrow intervals, proceeds as follows:

- Let $A^I x = b^I$
- Suppose A^c is the centre value of A^I
- Compute $B = (A^c)^{-1}$
- Use interval arithmetic to find BA^I and Bb^I
- Solve $(BA^I)\, x = (Bb^I)$ using for instance Gauss' elimination.

The error in the computed values of x^I is of the order of the square of the width of the coefficients' set [noted $O(W^2)$]. Miller (1972) provides us with an example, viz.

$$A^I = \begin{bmatrix} [2 \pm 0.1] & [1 \pm 0.1] \\ [1 \pm 0.1] & [1 \pm 0.1] \end{bmatrix}, \qquad b^I = \begin{bmatrix} [3 \pm 0.1] \\ [2 \pm 0.1] \end{bmatrix}$$

The set X is found from

$$\begin{aligned} 19x_1 + 9x_2 &\leqq 31\,, & 21x_1 + 11x_2 &\geqq 29 \\ 9x_1 + 9x_2 &\leqq 21\,, & 11x_1 + 11x_2 &\geqq 19 \end{aligned}$$

and is represented by a quadrilateral with vertices (1/3, 2), (1, 8/11), (17/11, 2/11) and (1, 4/3). Therefrom, the exact bound $[x]$ is given by

$$[x] = ([1/3, 17/11], [2/11, 2])^T$$

Now, multiplying by B which is given by

$$B = \begin{bmatrix} 2 & 1 \\ 1 & 1 \end{bmatrix}^{-1} = \begin{bmatrix} 1 & -1 \\ -1 & 2 \end{bmatrix}$$

we obtain

$$BA^I = \begin{bmatrix} [1 \pm 0.2] & [\pm 0.2] \\ [\pm 0.3] & [1 \pm 0.3] \end{bmatrix}, \qquad Bb^I = \begin{bmatrix} [1 \pm 0.2] \\ [1 \pm 0.3] \end{bmatrix}$$

The set $Y = \{y\colon BAy = Bb,\ BA \in BA^I,\ Bb \in Bb^I\}$ is obtained like before. It is represented by the quadrilateral with corners (1/3, 2), (3/5, 2/5), (17/11, 2/11) and (11/5, 14/5). Whence we get

$$[y_1] = [1/3, 11/5]\,, \qquad [y_2] = [2/11, 14/5]$$

But by applying Gauss' elimination, or the Gauss-Jordan variant of it, we can solve

$$BA^I z = Bb^I$$

for z, which is in turn given by

$$[z_1] = [1/5, 11/5], \qquad [z_2] = [2/11, 14/5]$$

The author later reworked the example using ε instead of 0.1 to estimate the order of the error between z^I and x^I. He found that

$$[x_1] = \left[\frac{1-7\varepsilon}{1-\varepsilon}, \frac{1+7\varepsilon}{1+\varepsilon}\right], \qquad [x_2] = \left[\frac{1-8\varepsilon}{1+\varepsilon}, \frac{1+8\varepsilon}{1-\varepsilon}\right]$$

and

$$[z_1] = \left[\frac{1-9\varepsilon+2\varepsilon^2}{1-3\varepsilon-10\varepsilon^2}, \frac{1+\varepsilon}{1-5\varepsilon}\right], \qquad [z_2] = \left[\frac{1-8\varepsilon}{1+\varepsilon}, \frac{1+4\varepsilon}{1-5\varepsilon}\right]$$

Here, note that $x^I \subseteq z^I$. Both $w(x^I)$ and $w(z^I)$ are of the form given by $12\varepsilon + 0(\varepsilon^2)$. The fact that z^I exceeds x^I by only $0(W^2)$, where W is equal to $w(A^I, b^I)$, is the main result obtained by the author. His proof is based on the lemma

$$\frac{1}{2} w([x_k]) = \sum_{i=1}^{n} \sum_{j=1}^{n} |q_{ki} x_j \varepsilon_{ij}| + \sum_{i=1}^{n} |q_{ki} \varepsilon_i| + 0(W^2)$$

which is a variant of Kuperman's result (cf. Sect. 1.3). Here q_{ki} is the (kth, ith) element of B. ε_{ij} is the error in a_{ij}, i.e.

$$[a_{ij}] = [a_{ij}^c \pm \varepsilon_{ij}]$$

ε_i is the error in b_i, i.e.

$$[b_i] = [b_i^c \pm \varepsilon_i]$$

x_j is the jth component of x^c, i.e. the solution to $A^c x = b^c$. Now solving, by interval arithmetic, $BA^I z = Bb^I$ must entail no further loss of accuracy. Incidentally, Gauss' method transforms BA^I into a diagonal matrix D. Thus, solving $Dz = d$ ensures that $x^I \subseteq z^I$ and that the same order of accuracy $O(W^2)$ is maintained. Further methods for estimating x^I can be found in Hansen and Smith (1967).

The two methods of Hansen discussed above are adequately implementable, especially with narrow intervals. For excessively large values of $w(A^I, b^I)$, however, more pronounced errors might appear in the results. It would have been more advantageous then to find a way to approach as much as possible the end corners of x^I before applying the above listed procedures. This was proposed by Hansen (1969). One can thence determine x^I very accurately, without the need for either estimating $(A^I)^{-1}$ or solving $BA^I z = Bb^I$ using interval arithmetic. In this case, even if $W = w(A^I, b^I)$ is large, the error in $w(x^I)$ would only depend on the round-offs, i.e. on the machine precision.

This method primarily involves solving $Ax = b$,with A and b fixed and uniquely chosen from A^I and b^I in such a way as to maximize or minimize x, thus obtaining

$x^I = x_{max} - x_{min}$. If $\dfrac{\partial x_k}{\partial a_{ij}}$ and $\dfrac{\partial x_k}{\partial b_i}$ are negative in sign, then the choice $\bar{a}_{ij}$ and $\bar{b}_i$ will make us approach x_{min}. In case the sign is reversed, the opposite choice should be made. As for the partial derivatives, they are simply obtained from the relations

$$A \frac{\partial x}{\partial a_{ij}} + \frac{\partial A}{\partial a_{ij}} x = 0$$

and

$$A \frac{\partial x}{\partial b_j} = \frac{\partial b}{\partial b_j}$$

To exemplify the technique, let us apply it to the example first listed above, namely

$$\begin{bmatrix} [2, 3] & [0, 1] \\ [1, 2] & [2, 3] \end{bmatrix} \begin{bmatrix} x_1 \\ x_2 \end{bmatrix} = \begin{bmatrix} [0, 120] \\ [60, 240] \end{bmatrix}$$

Now supposing we wish to determine $x_{1\,max}$ (in the fourth quadrant, with $x_1 \geqq 0$, $x_2 \leqq 0$), we will use A^c to represent A

$$A^c = \begin{bmatrix} 5/2 & 1/2 \\ 3/2 & 5/2 \end{bmatrix}, \qquad (A^c)^{-1} = \begin{bmatrix} 5/11 & -1/11 \\ -3/11 & 5/11 \end{bmatrix}$$

and

$$\frac{\partial x_1}{\partial a_{11}} = -[5/11 \quad -1/11] \begin{bmatrix} 1 & 0 \\ 0 & 0 \end{bmatrix} \begin{bmatrix} + \\ - \end{bmatrix} < 0$$

$$\frac{\partial x_1}{\partial a_{12}} = -[5/11 \quad -1/11] \begin{bmatrix} 0 & 1 \\ 0 & 0 \end{bmatrix} \begin{bmatrix} + \\ - \end{bmatrix} > 0$$

$$\frac{\partial x_1}{\partial a_{21}} = -[5/11 \quad -1/11] \begin{bmatrix} 0 & 0 \\ 1 & 0 \end{bmatrix} \begin{bmatrix} + \\ - \end{bmatrix} > 0$$

$$\frac{\partial x_1}{\partial a_{22}} = -[5/11 \quad -1/11] \begin{bmatrix} 0 & 0 \\ 0 & 1 \end{bmatrix} \begin{bmatrix} + \\ - \end{bmatrix} < 0$$

and

$$\frac{\partial x_1}{\partial b_1} = [5/11 \quad -1/11] \begin{bmatrix} 1 \\ 0 \end{bmatrix} > 0$$

$$\frac{\partial x_1}{\partial b_2} = [5/11 \quad -1/11] \begin{bmatrix} 0 \\ 1 \end{bmatrix} < 0$$

whence $x_{1\,\max}$ is obtained by solving

$$\begin{bmatrix} 2 & 1 \\ 2 & 2 \end{bmatrix} \begin{bmatrix} x_1 \\ x_2 \end{bmatrix} = \begin{bmatrix} 120 \\ 60 \end{bmatrix}$$

giving

$$x_{1\,\max} = 90$$

which is corroborated by Fig. 1. Carrying out the same calculations for $x_{1\,\min}$, $x_{2\,\max}$ and $x_{2\,\min}$, we obtain

$$x_{1\,\min} = -120$$
$$x_{2\max} = 240$$
$$x_{2\,\min} = -60$$

and the interval vector x^I is given by

$$x^I = ([-120, 90], [-60, 240])^T$$

Note that if any of the derivatives is null, its corresponding element a_{ij} is left as it is in the form of an interval $[a_{ij}]$. The previous methods are thus considered vital; they are used even when all derivatives are non-zero as in the previous example, to account for rounding errors. The error in $w(x^I)$ has now been improved; at best it is of the order $\varepsilon(l)$, l being the machine precision.

In practice, both A^c and b^c are usually given along with the range of the uncertainties in their elements. If x^I is the solution — computed in interval form $x^I = [\underline{x}, \bar{x}]$ by some method of interval arithmetic — then the error in x^c due to uncertainties in the data can be measured by the quantity

$$\frac{1}{2} w(x^I) = \frac{1}{2} \max_i (\bar{x}_i - \underline{x}_i)$$

We can further take as a measure of error, the interval metric

$$q(x^I, x^c) = \max \{\|\bar{x} - x^c\|, \|\underline{x} - x^c\|\} = \max_i \{\max (|\bar{x}_i - x_i^c|, |\underline{x}_i - x_i^c|)\}$$

This latter measure is sometimes preferred, on the grounds that x^c does not usually divide $\bar{x} - \underline{x}$ into two equal segments.

2.4 Method of Linear Programming

Linear programming facilities can provide us with a second approach for solving $A^I x = b^I$. This line of thought was adopted by Oettli and Prager (1964) and Oettli (1965). First, the authors set out to define the set

$$X = \{x \colon Ax = b, A \in A^I, b \in b^I\}$$

If x is a point belonging to the solution set X, then it must satisfy the following relation

$$\left|\sum_j a_{ij}^c x_j - b_i^c\right| \leq \sum_j \Delta a_{ij} |x_j| + \Delta b_i; \qquad i = 1, \ldots, n$$

Here, Δa_{ij} and Δb_i are the uncertainties in the data, i.e.

$$a_{ij}^c - \Delta a_{ij} \leqq a_{ij} \leqq a_{ij}^c + \Delta a_{ij}$$

and

$$b_i^c - \Delta b_i \leqq b_i \leqq b_i^c + \Delta b_i$$

To prove the above relation, write

$$(A^c + \delta A)\, x = b^c + \delta b$$

or alternatively

$$b^c - A^c x = \delta A x - \delta b$$

then

$$b_i^c - \sum_j a_{ij}^c x_j = \sum_j \delta a_{ij} x_j - \delta b_i; \qquad i = 1, \ldots, n$$

Now

$$|\delta a_{ij}| \leqq \Delta a_{ij} \quad \text{and} \quad |\delta b_i| \leqq \Delta b_i$$

Then the right-hand side certainly does not lie outside the intervals

$$\left[-\sum_j \Delta a_{ij} |x_j| - \Delta b_i, \sum_j \Delta a_{ij} |x_j| + \Delta b_i\right] \qquad i = 1, \ldots, n$$

Then for x to be an admissible solution of $A^I x = b^I$, the left-hand side must also lie within the same intervals, i.e.

$$b_i^c - \sum_j a_{ij}^c x_j \geq -\sum_j \Delta a_{ij} |x_j| - \Delta b_i; \qquad i = 1, \ldots, n$$

and

$$b_i^c - \sum_j a_{ij}^c x_j \leq \sum_j \Delta a_{ij} |x_j| + \Delta b_i; \qquad i = 1, \ldots, n$$

from which Oettli's inequality follows. A simple proof was further suggested by Hansen (1969). The fact that X denotes the set of all possible solutions x of

$Ax = b$ (A and b possibly assuming any of the values of A^I and b^I respectively) necessitates that

$$A^I x \cap b^I \neq \emptyset$$

Now writing $A^I x$ as $[\underline{A}x, \bar{A}x]$ for $x \geqq 0$, and b^I as $[\underline{b}, \bar{b}]$, we can conclude that for the intersection of $A^I x$ and b^I to be non-empty we must have,

$$\underline{A}x \leqq \bar{b} \quad \text{and} \quad \bar{A}x \geqq \underline{b}$$

Simultaneously, putting

$$\underline{A} = A^c - \Delta A\,, \qquad \bar{A} = A^c + \Delta A$$
$$\underline{b} = b^c - \Delta b\,, \qquad \bar{b} = b^c + \Delta b$$

results, for $x \geqq 0$, in the two inequalities

$$A^c x - b^c \leqq \Delta A x + \Delta b$$
$$A^c x - b^c \geqq -\Delta A x - \Delta b$$

wherefrom Oettli's compatibility condition directly follows. Now, we can utilize linear programming to maximize (or minimize) x, subject to the two above inequalities. This would yield extreme values located at some of the vertices of the space polyhedron. If $\bar{x}$ and $\underline{x}$ are respectively the maximum and minimum values of the objective function x processed with the simplex, then

$$x^I = [\underline{x}, \bar{x}]\,.$$

The reader is asked to apply the simplex on Hansen's example (1969) discussed previously, that is to maximize x subject to the inequalities

$$2x_1 \leqq 120\,, \qquad x_1 + 2x_2 \leqq 240$$
$$3x_1 + x_2 \geqq 0\,, \qquad 2x_1 + 3x_2 \geqq 60$$

For other ways of determining the solution set of $A^I x = b^I$, the reader may consult Hartfiel (1980).

As we have already pointed out, one of the problems usually enountered when using this approach is the nonconvexity of the region containing the possible solutions. Fortunately enough, in practice Δa_{ij} is small and the polyhedron lies in just one orthant. This allows for the simple application of linear programming. However, if Δa_{ij} is large, Cope and Rust (1979) argue, application of the same technique might still be possible. In fact, whilst Oettli and his co-workers require that Δa_{ij} be small, Cope and Rust allow the solution set to spread over many orthants, and even to become unbounded. They look for bounds on all of the solutions lying in any one orthant. A word for the reader wishing to compare this line of methods to that of Hansen's for determining $(A^I)^{-1}$: linear programming is much slower, but all the more accurate — especially with wide intervals.

However, one must bear in mind the fact that x^I will not satisfy the equations $A^I x^I = b^I$, be they computed using Hansen's method, linear programming or any other method. This is why we write the equations in the form $A^I x = b^I$, where x is any element of the set X given by

$$X = \{x \colon Ax = b,\ A \in A^I,\ b \in b^I\}$$

As for x^I — the minimum interval enclosing X — it satisfies the relation

$$A^I x^I \supseteq b^I$$

This can be seen from Hansen's principle $A^I x^i \cap b^I \neq \emptyset$, for an $x^i \in X \in [\underline{x}, \overline{x}]$. In other words, one can conceive that

$$\bigcup_i A^I x^i \supseteq b^I$$

(see also exercise 2.9), Figure 2 illustrates this fact.

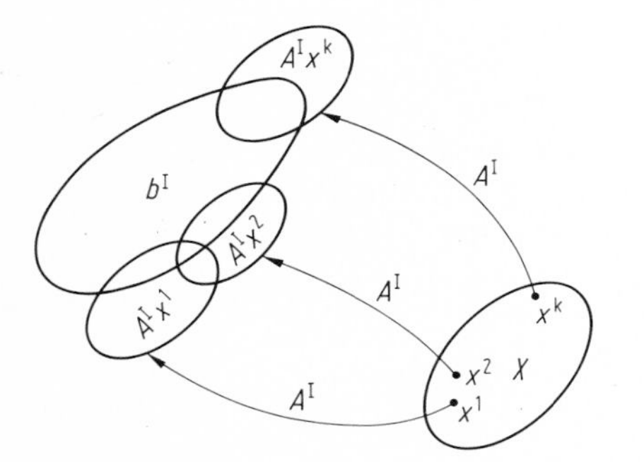

Fig. 2.2

Fig. 2.3

In general, the big the area of X enclosed in x^I, the tighter the bound $A^I x^I \supseteq b^I$. For example, for the equations

$$[2, 4]\, x_1 + [1, 2]\, x_2 = [-1, 1]$$
$$[1, 2]\, x_1 + [3, 4]\, x_2 = [-1, 1]$$

sketched in Fig. 3, X represents 70% of the surface area of x^I.

For the famous example in Barth and Nuding (1974) depicted in Fig. 4 and stated as

$$[2, 4]\, x_1 + [-2, 1]\, x_2 = [-2, 2]$$
$$[-1, 2]\, x_1 + [2, 4]\, x_2 = [-2, 2]$$

the area of the set X is relatively small compared to that of x^I. The difference between both areas is called the *overestimation error*. In practice, different algorithms produce different x^I's and accordingly different overestimation errors. We have seen that Gauss' method, for example, produces large overestimation errors in comparison with other methods. In fact, one of numerical analysts major concerns is presently to find new techniques for the reduction of this error. Their goal is to find bound-conserving algorithms, by contrast with stable ones as stated in standard real analysis. In this respect, two choices were possible: finding classes of data A, b for which Gauss' method is bound-conserving (as for the class of M matrices discussed in Barth and Nuding (1974) and in Beeck (1974)); or resorting to iterative methods to improve on the results as we shall see later on.

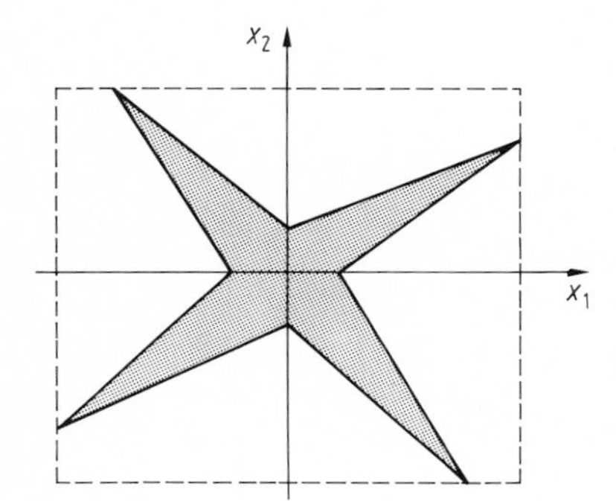

Fig. 2.4

2.5 A-Posteriori Bounds

Error analysis of the solution of the equation $Ax = b$ usually involves three phases:

1. Design of a technique to be stable and virtually insensitive to the uncertainties in the coefficients of A and b. In any case, the solution will contain a certain error due to the machine precision, which can be forecast using Wilkinson's a-priori analysis.
2. After having solved the equations, finding a way to judge of the accuracy of the results; a procedure aiming at deciding whether to accept or reject the results and termed a-posteriori analysis. Results are judged accurate or not usually by comparing the residual vector $r = A\hat{x} - b$ to the allowable uncertainties in A and b: if these latters are allowed to vary in a limited range, then r is supposed not to exceed a certain bound.
3. Investigating the possibility of improving on the results, although they might be acceptable on an a-posteriori basis. This can be achieved using iteration; its discussion is postponed to the next chapter.

As the term implies, a-posteriori analysis yields a prognosis of the operation executed on the problem $Ax = b$. We already knew of the uncertainties in A and b before running our algorithm, be they due to limited precision of the machine in representing the numbers, or to an incapacity on our behalf to determine our data accurately. Such uncertainties are not defects in themselves, for they will always exist. Rather the defect might lie in obtaining results that are incompatible with the amount of anticipated uncertainty. Analyzing how good the results are in relation to how precise our data were is what we term *a-posteriori analysis*.

Oettli's compatibility condition for the existence of a solution x that satisfies $A^I x = b^I$ — as stated before — has inspired Rigal and Gaches (1967) to use it also as a test, for a linear system, of the compatibility of a given solution with the data. Despite their having fixed values, both A and b still suffer from inaccuracies, either inherent or introduced to the system. If $\hat{x}$ is an approximate solution of the system $Ax = b$, then it is considered compatible with the linear system's data if it satisfies

$$(A + \delta A)\,\hat{x} = b + \delta b$$
$$N_1(\delta A) \leqq \alpha$$
$$N_2(\delta b) \leqq \beta$$

where δA and δb are matrix and vector of uncertainties. N_1 and N_2 are some definitions of norm, and α and β are measures of the uncertainties. The decision whether to accept or reject $\hat{x}$ according to its being or not compatible with the uncertainties in the data depends on us finding a class of matrices δA and vectors δb, bounded by the above described measures called the *loci of indiscernible data*, that satisfy the perturbed problem. The condition of compatibility is then found to be

$$\|A\hat{x} - b\|_p \leqq \alpha\|\hat{x}\|_q + \beta$$

with $1/p + 1/q = 1, p \geqq 1$. To obtain this fine result, we can use as before the equation

$$A\hat{x} - b = \delta b - \delta A\hat{x}$$

put in Oettli's form, since the intervals of uncertainty in A and b are $[-\Delta A, \Delta A]$ and $[-\Delta b, \Delta b]$, that is

$$\Big|\sum_j a_{ij}\hat{x}_j - b_i\Big| \leq \sum_j \Delta a_{ij}\,|\hat{x}_j| + \Delta b_i; \qquad i = 1, \ldots, n\,.$$

By summing over all values of i, we would obtain

$$\sum_i \Big|\sum_j a_{ij}\hat{x}_j - b_i\Big| \leq \Big(\sum_{i,j} \Delta a_{ij}\Big) \max_i |\hat{x}_i| + \sum_i \Delta b_i$$

which is Rigal and Gaches' result for $p = 1$, provided we define

$$\|x\|_p = \Big(\sum_i |x_i|^p\Big)^{1/p}$$

$$\|A\|_p = \Big(\sum_{i,j} |a_{ij}|^p\Big)^{1/p}$$

For a general value of p greater than unity, the result would still be valid if we raise both sides of Oettli's inequality to the power p and sum again over all values of i. Then taking the pth root, we obtain

$$\begin{aligned}\|A\hat{x} - b\|_p &\leq \Big(\sum_i \Big(\sum_j \Delta a_{ij} \,|\hat{x}_j| + \Delta b_i\Big)^p\Big)^{1/p} \\ &= \|\Delta A\,|\hat{x}| + \Delta b\|_p \\ &\leqq \|\Delta A\,|\hat{x}|\|_p + \|\Delta b\|_p \\ &\leqq \|\Delta A\|_p \,\|\hat{x}\|_q + \|\Delta b\|_p\,,\end{aligned}$$

the exact result required. In fact, the authors' main contribution lies in the remark that the expression

$$\|\Delta A\hat{x}\|_p \leqq \|\Delta A\|_p \,\|\hat{x}\|_q$$

sets a tighter bound than both the subordinate Hölder matrix and consistent vector norms. This becomes clear when we take $\Delta A = yz^*$. The angles between $\hat{x}$ and all vectors $[\Delta a_{i1}, \Delta a_{i2}, \ldots, \Delta a_{in}]$; $i = 1, \ldots, n$ become equal. By applying the idea of dual norm discussed in Sect. 1.2, we obtain

$$\sup_{x\neq 0} \frac{\|\Delta Ax\|_p}{\|x\|_q} = \sup_{x\neq 0} \frac{\|yz^*x\|_p}{\|x\|_q} = \|y\|_p \sup_{x\neq 0} \frac{|z^*x|}{\|x\|_q} = \|y\|_p \,\|z\|_p = \|\Delta A\|_p$$

As for the calculation of δA and δb required for error adjustment in the solution $\hat{x}$, the authors suggested

$$\delta b = \frac{\beta r}{\alpha \,\|\hat{x}\|_q + \beta}, \qquad r = A\hat{x} - b$$

and $\delta A = yz^*$, where z is a vector satisfying the relation $z^*\hat{x} = 1$. Therefore, from $r = A\hat{x} - b = \delta b - \delta A\hat{x} = \delta b - yz^*\hat{x}$, $y = \delta b - r$. Then

$$\begin{aligned}\delta A = yz^* &= (\delta b - r)\, z^* \\ &= -\frac{\alpha \,\|\hat{x}\|_q}{\alpha \,\|\hat{x}\|_q + \beta} rz^*\end{aligned}$$

The reader may refer to Oettli, Prager and Wilkinson (1965) for similar ideas. Although the compatibility condition derived by Rigal and Gaches was meant mainly for square matrices, it is still valid for least-squares problems. For a discussion of this topic, the reader may refer to Kovarik (1977).

Since in the above analysis, it is only required to find one possible δA, we have chosen the simplest one $\delta A = yz^*$. Such perturbation is called *perturbation of rank one*. Perturbations of finite ranks, beside setting tighter bounds as was seen above, provide a representation of δx in the same form as the perturbations in the data (see exercise 5.1 and also Rall (1979)).

Unlike *forward error estimation* in chapter one which starts from assumptions about data perturbations and obtains a comparison between the actual solution and the true solution, *backward error estimation* as developed here, assumes the solution obtained to be the exact solution of some perturbed problem and proceeds to estimate the corresponding changes in the data.

To demonstrate the above condition on an example, let us consider the following, executing our computations on a 10-digit machine with floating-point arithmetic:

$$A = \begin{bmatrix} 1 & 1 \\ 1 & 4 \end{bmatrix}, \qquad b = \begin{bmatrix} 2 \\ 1 \end{bmatrix}$$

For a solution $\hat{x} = (2.333\,333\,333, -0.333\,333\,333)^T$, the residual is $r = A\hat{x} - b = (0, 10^{-9})^T$. Now, taking $p = 1$, then

$$\|\Delta A\|_1 = (5)\,(10^{-10})\,(1 + 1 + 1 + 4) = (35)\,(10^{-10})$$
$$\|\Delta b\|_1 = (5)\,(10^{-10})\,(2 + 1) = (15)\,(10^{-10})$$

and

$$\|r\|_1 = 10^{-9} \leqq \|\Delta A\|_1\,\|\hat{x}\|_\infty + \|\Delta b\|_1 = (35)\,(10^{-10})\,(2.333\,333\,333) + (15)\,(10^{-10})$$

Hence, $\hat{x}$ is a realistic solution despite the round-off errors. Also

$$\delta b = \frac{(15)\,(10^{-10})}{(35)\,(10^{-10})\,(2.33\,\ldots) + (15)\,(10^{-10})}\,(0, 10^{-9})^T < (0, (2)\,(10^{-10}))^T$$

and

$$\delta A = \left(\begin{bmatrix} 0 \\ (2)\,(10^{-10}) \end{bmatrix} - \begin{bmatrix} 0 \\ 10^{-9} \end{bmatrix} \right) z^*$$

where z can be chosen as $z = (\frac{1}{2}, \frac{1}{2})^T$, satisfying $z^T x = 1$ and also $\|z\| \cdot \|\delta b - r\| \leqq \alpha$. δA can be chosen as

$$\delta A = -\begin{bmatrix} 0 & 0 \\ (4)\,(10^{-10}) & (4)\,(10^{-10}) \end{bmatrix}$$

satisfying the perturbed problem

$$\left(\begin{bmatrix} 1 & 1 \\ 1 & 4 \end{bmatrix} - \begin{bmatrix} 0 & 0 \\ (4)\,(10^{-10}) & (4)\,(10^{-10}) \end{bmatrix} \right) \begin{bmatrix} 2.333\,333\,333 \\ -0.333\,333\,333 \end{bmatrix} = \begin{bmatrix} 2 \\ 1 \end{bmatrix} + \begin{bmatrix} 0 \\ (2)\,(10^{-10}) \end{bmatrix}$$

with exactitude, up to the accuracy of a ten-digit mantissa. Note that δA and δb are not unique, but depend on the particular choice of α and β.

It is also worth noting that since Rigal and Gaches compatibility condition, as stated above, is just another version of Oettli's criterion, the latter can still be used in its raw form, being that form suggested by Skeel (1979). If r satisfies

$$|r| \leqq \Delta A\ |\hat{x}| + \Delta b$$

where $|r|$ stands for the vector of absolute values, then $\hat{x}$ is a realistic solution. Then for a machine of precision ε, $\Delta A = \varepsilon\ |A|$ and also $\Delta b = \varepsilon\ |b|$, and

$$|r| \leqq \varepsilon(|A|\ |\hat{x}| + |b|)$$

becomes another form of the above criterion when testing the compatibility of $\hat{x}$ with the error ε in the data using floating-point arithmetic. δA and δb are then chosen as

$$\delta b = -H|b|$$
$$\delta A = H|A|\ \mathrm{diag}\ (\mathrm{sgn}\ \hat{x})$$

where H is a diagonal matrix, with $|h_{ii}| \leqq \varepsilon$. For the above example, δb and δA can be chosen to fit into the perturbed problem

$$\left(\begin{bmatrix} 1 & 1 \\ 1 & 4 \end{bmatrix} + \begin{bmatrix} \eta_1 & -\eta_1 \\ \eta_2 & -4\eta_2 \end{bmatrix}\right) \begin{bmatrix} 2.333\,333\,333 \\ -0.333\,333\,333 \end{bmatrix} = \begin{bmatrix} 2 - 2\eta_1 \\ 1 - \eta_2 \end{bmatrix}$$

with $\eta_1 = 0, |\eta_2 = -2.142\,857\,144 \times 10^{-10}| < \varepsilon = 5 \times 10^{-10}$.

The above compatibility criterion has been widely implemented in many packages. For example in the IMSL one, if $\hat{x}$ is a computed solution of $Ax = b$, then it is considered compatible with the uncertainties in the data as a result of round-offs, if the machine accuracy ε is greater than p, where

$$p = \max_{1 \le i \le n} \frac{\left| b_i - \sum_{j=1}^{n} a_{ij}\hat{x}_j \right|}{BN + AN \sum_{j=1}^{n} |\hat{x}_j|}$$

and where

$$BN = \max_{1 \le i \le n} |b_i|, \qquad AN = \max_{1 \le i, j \le n} |a_{ij}|.$$

The above compatibility condition can also be used to control the value of r. For example, by scaling the equations into the form

$$|Dr| \leqq \varepsilon |D|\ (|A|\ |\hat{x}| + |b|)$$

where D is a scaling diagonal matrix. One can choose

$$d_{ii} = \frac{1}{|a_{i1}|\ |\hat{x}_1| + \ldots + |a_{in}|\ |\hat{x}_n| + |b_i|}$$

to bound $|Dr|$ by $\varepsilon(1, 1, \ldots, 1)^T$. However, we should not be misled by the idea that we have reached an optimum solution, since $\|\delta x\|$ is bounded by $\| |A^{-1}| |r| \| \leqq \varepsilon \|A^{-1}D^{-1}\|$ which could assume a high value. If for instance in the example considered in Sect. 1.5

$$\begin{bmatrix} 1/2 & 1/3 & 1/4 \\ 1/3 & 1/4 & 1/5 \\ 1/4 & 1/5 & 1/6 \end{bmatrix} \begin{bmatrix} x_1 \\ x_2 \\ x_3 \end{bmatrix} = \begin{bmatrix} 1 \\ 1 \\ 1 \end{bmatrix}$$

we choose $D = \operatorname{diag}(1/42, 1/32, 1/26)$, we get for $\hat{x}$:

$$\hat{x} = (12.000\,000\,14, -60.000\,000\,39, 60.000\,000\,23)^T$$

which is a worse result than when the system was first scaled as seen. Yet, the new residual r is such that

$$|r| < \varepsilon(1, 1, \ldots, 1)^T$$

This explains why there is no algorithm that performs satisfactorily in scaling any general matrix. However, the ratio

$$\frac{\max(|A|\,|\hat{x}| + |b|)}{\min(|A|\,|\hat{x}| + |b|)}$$

can give an indication of how poor is the scaling of a system.

In the above analysis, the authors were not concerned with identifying the domain of compatible solutions. Rather, their concern lay in making sure that the computed solution $\hat{x}$ lies within this domain, which would be termed realistic solution in case it satisfies this criterion. Following another methodolgy, other authors have devised an expensive way of defining the domain of uncertainty of $\hat{x}$, thus determining the number of significant digits in each of the unknowns as well. This was done statistically, as in the work of LaPorte and Vignes (1975). If the residual r_i of each equation $\sum_j a_{ij}\hat{x}_j - b_i$ is calculated and found to satisfy the condition of normed residuals, namely

$$\varrho_i = \frac{|r_i|}{\hat{r}_i} \leqq 1, \qquad i = 1, \ldots, n$$

with

$$\hat{r}_i = 2^{-N} n \sqrt{\sum_{j=1}^{n} (a_{ij}\hat{x}_j)^2 + b_i^2}$$

where N is the number of bits of binary mantissa, we call the solution $\hat{x}$ *solution informatique*. All of such informative solutions constitute in $\mathbb{R}^n$ a set $\mathscr{D}$ which is termed *domaine d'incertitude*, that is

$$\mathscr{D} = \{\hat{x}^{(1)}, \hat{x}^{(2)}, \ldots, \hat{x}^{(k)}\}.$$

Every point in $\mathscr{D}$ corresponds to an informative solution of the system, and the bounds of this domain permit a definition of the uncertainty in each of the unknowns. If $\bar{x}_i$ and δ_i^2 are respectively the mean and variance of the unknown of order i in the population of $\mathscr{D}$, then

$$\hat{\varepsilon}_i = \sqrt{(\hat{x}_i^{(1)} - \bar{x}_i)^2 + \delta_i^2}$$

Knowing $\hat{\varepsilon}_i$, we can estimate the number of significant decimal digits c_i of each unknown from the relation

$$\hat{\varepsilon}_i = |\hat{x}_i^{(1)}| \times 10^{-c_i}$$

This relation characterizes the relative error in each of the unknowns. As for defining $\mathscr{D}$, the authors have used a method which they termed *méthode de permutation-perturbation*. Obviously, from its very name, the method consists in a permutation of the columns of A together with a perturbation of its elements, with a solution each time for $\hat{x}$.

2.6 Two Problems in Interval Analysis

This section will deal specifically with the following two problems, find a

1. Maximal $x^I \subseteq X = \{x : A^I x \subseteq b^I\}$
2. Minimal $x^I \supseteq X = \{x : A^I x \supseteq b^I\}$

The reason for the assignment of a separate section for their discussion their apparition in a variety of applications. Of the latter, we mention the domain of tolerance analysis. To clarify, we take the precise example of an electrical filter having an external behaviour R which we call response. This latter is obviously a function of the filter's elements as well as frequency, i.e.

$$R = R(x_1, x_2, \dots, x_n, \omega)$$

Now suppose that due to some discrepancies in the elements resulting from ageing, temperature changes, manufacturing defects, etc. ..., the actual response $\tilde{R}(\hat{x}_1, \hat{x}_2, \dots, \hat{x}_n, \omega_i)$ will differ at each frequency ω_i from the ideal $R(x_1, \dots, x_n(\omega_i)$. Then if we set a bound for the variation of $\tilde{R}$ from R at each frequency ω_i, the question will arise as to the maximum allowable deviation (in plus or minus) in each of the elements $x_j (j = 1, \dots, n)$ such that $\tilde{R}$ deviates from the ideal R by no more than a prescribed amount $\pm\Delta R(\omega_i)$. The problem can be formulated otherwise as follows: Find the maximal Δx^I, such that $\tilde{R} - R \subseteq \Delta R$. For further elaboration, we define a sensitivity matrix S as

$$S: s_{ij} = \left.\frac{\partial R}{\partial x_j}\right|_{\omega_i}$$

and set as a condition that for small perturbations

$$-\Delta R \leqq \tilde{R} - R = S\,\Delta x \leqq \Delta R$$

where $\Delta x^T = (\tilde{x}_1 - x_1, \ldots, \tilde{x}_n - x_n)$, $\Delta R^T = (\Delta R(\omega_1), \ldots, \Delta R(\omega_m))$, n and m being respectively the number of elements and the number of sample frequency points. In general, we have that $m \geqq n$. The problem is now more defined and we look for the maximal $\Delta x^I \subseteq X = \{\Delta x\colon -\Delta R \leqq S\,\Delta x \leqq \Delta R\}$. A more generalized form of the problem would be: Find the

$$\text{Maximal } x^I \subseteq X = \{x\colon A^I x \subseteq b^I\}$$

The practical example stated above becomes therefore a special case of this general form. Likewise in the second of the two problems, we basically handle the exactly opposite situation. When we search for the

$$\text{Minimal } x^I \supseteq X = \{x\colon A^I x \supseteq b^I\}$$

we are in fact looking for the minimum allowable x_i^I, $i = 1, \ldots, n$, such that $A^I x$ encloses still b^I.

The above two problems have not deserved much attention in the literature, and yet they have been touched upon briefly in the German school. For instance, in Nuding and Wilhelm (1972), the authors called the solution respectively of $A^I x \subseteq b^I$ and $A^I x \supseteq b^I$ *innere Lösung* and *äußere Lösung* and forwarded a simple example of crane's design to which these concepts can be applied.

It is obvious that both problems differ from Hansen's problem stated as to find a minimal $x^I \supseteq X = \{x\colon A^I x = b^I\}$. The set X in Hansen's problem is larger than its counterpart in the two above problems, since $A^I x = b^I$ implies also $A^I x \cap b^I \neq \emptyset$, a special case containing both sets $A^I x \subseteq b^I$ and $A^I x \supseteq b^I$, when they assume respectively a maximum and a minimum value of x^I. However, unlike Hansen's problem which has always a solution, the above two problems may not have one. For consider

$$\begin{bmatrix} [1,2] & [-2,1] \\ [0,1] & [2,2] \end{bmatrix} \begin{bmatrix} x_1 \\ x_2 \end{bmatrix} \subseteq \begin{bmatrix} [0,10] \\ [10,20] \end{bmatrix}$$

having a region X, defined for $x_1, x_2 \geqq 0$ by

$$\begin{aligned} &x_1 - 2x_2 \geqq 0\,, \qquad 2x_1 + x_2 \leqq 10 \\ &2x_2 \geqq 10\,, \qquad x_1 + 2x_2 \leqq 20 \end{aligned}$$

which is empty. Same thing applies to all other quadrants. While for the same A^I and b^I, the set $X = \{x\colon A^I x \supseteq b^I\}$ has a solution $x = (10{,}5)^T$. We will now represent the two problems schematically, and derive conditions for the existence of a solution X.

For the first problem, illustrated in Fig. 5, we can assert that if

$$X_1 = \{x : \bar{A}x = b^I\}, \qquad X_2 = \{x : \underline{A}x = b^I\}$$

then $A^I x \subseteq b^I$ has a solution set X, only if

$$X_1 \cap X_2 \neq \emptyset$$

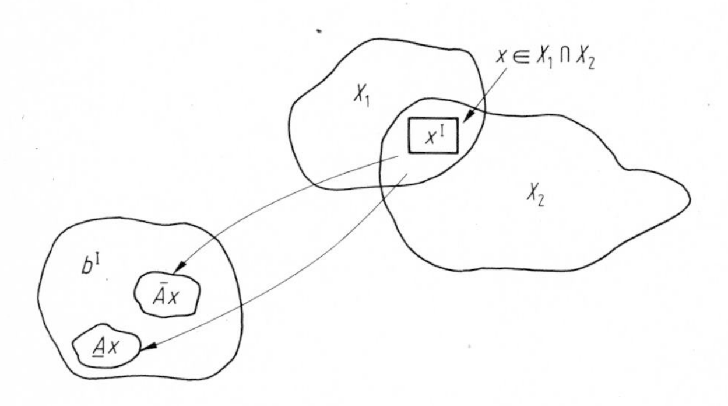

Fig. 2.5

The proof relies upon a description of the interval set $A^I x$ — valid for all x- communicated to the author by Dr. J. Rohn. $A^I x$ is defined by

$$A^I x = [A^c x - \Delta A|x|, A^c x + \Delta A|x|]$$

wherefrom applying for instance $A^I x \cap b^I \neq \emptyset$ leads directly to Oettli's criterion (cf. Sect. 2.4). Whereas for $A^I x \subseteq b^I$ to have a solution x, this implies that the set X described by the two inequalities

$$A^c x - \Delta A|x| \geqq \underline{b}, \qquad A^c x + \Delta A|x| \leqq \bar{b}$$

is nonempty. The same set can also be described by

$$\Delta A|x| - \Delta b \leqq A^c x - b^c \leqq -\Delta A|x| + \Delta b,$$

or even more briefly

$$|A^c x - b^c| \leqq -\Delta A|x| + \Delta b$$

But $-\Delta A|x| + \Delta b$ is less than both $-\Delta Ax + \Delta b$ and $\Delta Ax + \Delta b$. Similar argument applies to $\Delta A|x| - \Delta b$, to give

$$-\Delta Ax - \Delta b \leqq A^c x - b^c \leqq -\Delta Ax + \Delta b$$
$$\Delta Ax - \Delta b \leqq A^c x - b^c \leqq \Delta Ax + \Delta b$$

or that

$$-\Delta b \leqq \bar{A}x - b^c \leqq \Delta b$$
$$-\Delta b \leqq \underline{A}x - b^c \leqq \Delta b$$

which constitute a necessary condition for the equations $\bar{A}x = b^I$ and $\underline{A}x = b^I$ to have each a solution x (see Sect. 2.4). And since x satisfies simultaneously both of them, then $x \in X_1 \cap X_2$; completing thus the proof.

Now comes the difficult task of choosing x^I, one of maximal width. Unfortunately, no unique value of x^I can be generally found. For consider

$$\begin{bmatrix} 2 & 1 \\ 1 & 2 \end{bmatrix} \begin{bmatrix} x_1 \\ x_2 \end{bmatrix} = \begin{bmatrix} [-1, 1] \\ [-2, 2] \end{bmatrix}$$

Here, there is no need for setting as a condition that $Ax \subseteq b^I$, since it is automatically satisfied given that A = constant, wherefrom

$$Ax = b^I \Rightarrow Ax \cap b^I \neq \emptyset \Rightarrow Ax \subseteq b^I$$

The region X is a parallelogram of vertices (—4/3, 5/3), (0, 1), (4/3, —5/3) and (0, —1). Any maximum interval chosen inside X will satisfy the relation $Ax^I \subseteq b^I$. One can choose x^I as $x^I = ([-1/2, 0], [0, 1])^T$ to yield $Ax^I = ([-1, 1], [-1/2, 2])^T$, which is included in b^I. Other choices for x^I may well be $x^I = ([-1/4, 1/4], [-1/2, 1/2])^T$ and $x^I = ([0, 1/2], [-1, 0])^T$ ... etc. Hence no unique x^I exists. In practical applications, other conditions are usually imposed on x^I. For instance, due to some economic reasons $[x_1]$, $[x_2]$ ... etc. may be confined to a certain range, and so forth. But having that $A^I x \subseteq b^I$ as our starting assumption will at best enable us to find the extreme points of X. As to choosing x^I, it becomes the task for an engineer, who can judge which of the values of x^I inside X is the most favorable one.

To determine therefore the extreme points of X, one can opt for the linear programming approach (see Sect. 2.4) and having as constraints

$$\Delta A|x| - \Delta b \leqq A^c x - b^c \leqq -\Delta A|x| + \Delta b\,.$$

Another method which can be used, is when $\bar{A}x = b^I$ and $\underline{A}x = b^I$ define X. The two equations are written in the form

$$\begin{bmatrix} \bar{A} \\ -- \\ \underline{A} \end{bmatrix} x = \begin{bmatrix} b^I \\ --- \\ b^I \end{bmatrix}$$

or alternatively that $A_t x = b_t^I$, where t stands for total. The problem is now easier and the condition that $A_t x \cap b_t^I \neq \emptyset$ is directly obtainable. Note that this last contition is not trivial. It would have been so had A^t been square; i.e. if A was constant and such a condition would have been automatically satisfied. But because A_t is a rectangular matrix of order $2n \times n$, the equation $A_t x = b_t^I$ may not have a solution at all. It would have a solution only if $A_t A_t^i b_t = b_t$, $\forall\, b_t \in b_t^I$ (see

Sect. 4.1). In other words, a necessary condition for the existence of a set $X = \{x: A^I x \subseteq b^I\}$ is that $A_t A_t^i b_t^I \cap b_t^I \neq \emptyset$. The reader may exercise by checking the last condition on the two numerical examples above.

Consider the example in Sect. 2.2

$$[2, 3]\, x_1 + [0, 1]\, x_2 = [0, 120]$$
$$[1, 2]\, x_1 + [2, 3]\, x_2 = [60, 240]$$

then

$$A^I = \begin{bmatrix} [2, 3] & [0, 1] \\ [1, 2] & [2, 3] \end{bmatrix}, \qquad b^I = \begin{bmatrix} [0, 120] \\ [60, 240] \end{bmatrix}$$

$$A_t = \begin{bmatrix} 3 & 1 \\ 2 & 3 \\ 2 & 0 \\ 1 & 2 \end{bmatrix}, \qquad A_t^i = \begin{bmatrix} 0 & 0 & 1/2 & 0 \\ 0 & 1/3 & -1/3 & 0 \end{bmatrix}$$

and $x = A_t^i b$. Furthermore

$$A_t A_t^i = \begin{bmatrix} 0 & 1/3 & 7/6 & 0 \\ 0 & 1 & 0 & 0 \\ 0 & 0 & 1 & 0 \\ 0 & 2/3 & -1/6 & 0 \end{bmatrix}$$

and

$$A_t A_t^i b_t^I \cap b_t^I = \begin{bmatrix} [20, 220] \\ [60, 240] \\ [0, 120] \\ [20, 160] \end{bmatrix} \cap \begin{bmatrix} [0, 120] \\ [60, 240] \\ [0, 120] \\ [60, 240] \end{bmatrix} \neq \emptyset$$

Now, to determine the extreme points, we calculate $x_{1\,\max}$, $x_{2\,\max}$, $x_{1\,\min}$ and $x_{2\,\min}$. We will carry out these calculations for $x_{1\,\max}$ only. We have that

$$x_1 = \tfrac{1}{2} b_3$$

where $b^T = (b_1 \; b_2 \; b_3 \; b_4)$

From consistency, we get

$$\frac{1}{3} b_2 + \frac{7}{6} b_3 = [0, 120]$$

$$\frac{2}{3} b_2 - \frac{1}{6} b_3 = [60, 240]$$

with

$$b_2 \in [60, 240], \qquad b_3 \in [0, 120]$$

Therefore, to maximize x_1 is in fact to maximize b_3, subject to the last conditions. Using Hansen's last seen method, we get $b_3 = 72$, $b_2 = 108$. Hence solving the equations

$$2x_1 + 3x_2 = 108$$
$$2x_1 = 72$$

yields the point (36, 12) as shown in Fig. 6, and so forth.

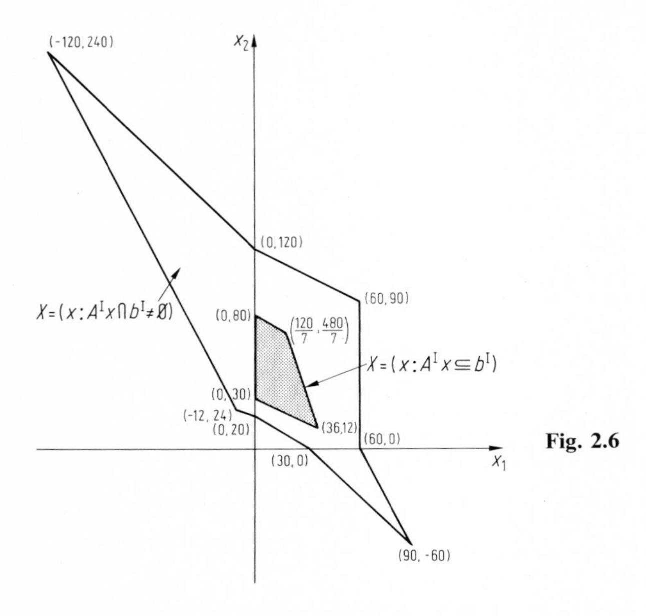

Fig. 2.6

The second problem, namely: find the

$$\text{Minimal } x^I \supseteq X = \{x \colon A^I x \supseteq b^I\}$$

follows a similar argument. Here again, X could be found empty for some A^I and b^I. For example, if

$$A^I = \begin{bmatrix} [1, 2] & [-1, 1] \\ [0, 1] & [2, 2] \end{bmatrix}, \qquad b^I = \begin{bmatrix} [0, 10] \\ [10, 20] \end{bmatrix}$$

then $A^I x \supseteq b^I$ defines a set X described for $x_1, x_2 \geqq 0$ by

$$x_1 - x_2 \leqq 0, \qquad 2x_1 + x_2 \geqq 10$$
$$2x_2 \leqq 10, \qquad x_1 + 2x_2 \geqq 20$$

which is void, and same applies to all other quadrants.

To find out whether a solution set X exists, such that

$$X = \{x: A^I x \supseteq b^I\},$$

let

$$X_1 = \{x: A^I x = \bar{b}\}, \qquad X_2 = \{x: A^I x = \underline{b}\}$$

then $A^I x \supseteq b^I$ has a solution x, iff

$$X_1 \cap X_2 \neq \emptyset$$

The proof follows the same lines as for the first problem, which we shall not repeat. Rather we shall be contented to define X. Again since x satisfies simultaneously

$$A^I x = \bar{b} \quad \text{and} \quad A^I x = \underline{b}$$

then, from Sect. 2.4, we have

$$-\Delta A\,|x| \leqq A^c x - \bar{b} \leqq \Delta A\,|x|$$
$$-\Delta A\,|x| \leqq A^c x - \underline{b} \leqq \Delta A\,|x|$$

or that x satisfies

$$-\Delta A\,|x| + \Delta b \leqq A^c x - b^c \leqq \Delta A\,|x| - \Delta b$$

or even more briefly, the set X is such that

$$|A^c x - b^c| \leqq \Delta A\,|x| - \Delta b$$

This is identical to $X = \{x: A^I x \supseteq b^I\}$.

At this point, one must note, that altough x could be empty, there always exists an x^I, such that $A^I x^I \supseteq b^I$. A corresponding situation is not to be encountered in the first problem, whereby an existence of a nonzero interval vector x^I such that $A^I x^I \subseteq b^I$ necessitates the existence of a set X whose points belong to x^I satisfying $A^I x \subseteq b^I$. The reason for which, in the second problem, there always exists an x^I such that $A^I x^I \supseteq b^I$, can be easily reached from Hansen's problem $A^I x = b^I$ having always as solution a nonempty set X and a minimal $x^I \supseteq X$ such that $A^I x^I \supseteq b^I$. In our case since the condition $A^I x = b^I$ is released, x^I is narrower than its corresponding one in Hansen's problem.

To search for a minimal x^I such that $A^I x^I \supseteq b^I$, when $A^I x \supseteq b^I$ has no solution set X, may be achieved using a similar method to Hansen's third method in Sect. 2.3. Let $x^1 \in X_1$ and $x^2 \in X_2$, where

$$X_1 = \{x: A^I x^1 = \bar{b} \text{ or } A^I x^1 \supseteq \bar{b}\}$$

and

$$X_2 = \{x: A^I x^2 = \underline{b} \text{ or } A^I x^2 \supseteq \underline{b}\}$$

It follows that

$$b^I \subseteq A^I x^1 \cup A^I x^2 \subseteq A^I x^I$$

where x^I is the interval whose vertices are x^1 and x^2, and by minimizing $\|x^1 - x^2\|_E$ gives the minimal x^I. Writing the last scalar function in the form

$$\|x^1 - x^2\|_E^2 = \sum_{i=1}^{n} (x_i^1 - x_i^2)^2$$

and then differentiating with respect to a_{ij}, and reckoning its sign (positive or negative), we will obtain the necessary hint as to which value of a_{ij}, $\|x^1 - x^2\|_E$ becomes minimum. This is a variant of Hansen's last method discussed before. If the sign of the expression

$$\frac{\partial \|x^1 - x^2\|_E^2}{\partial a_{ij}} = 2 \sum_{i=1}^{n} (x_i^1 - x_i^2) \left(\frac{\partial x_i^1}{\partial a_{ij}} \text{ or } \frac{-\partial x_i^2}{\partial a_{ij}} \right)$$

is positive, then $\underline{a}_{ij}$ will reduce $\|x^1 - x^2\|_E$; $\bar{a}_{ij}$ is the appropriate choice if the sign is negative. Note that the reason for writing the second bracket in the right-hand side of the expression the way it is, is that x^1 depends on a_{ij} in the problem $A^I x^1 = \bar{b}$, while x^2 does not in the same problem. Let us illustrate the above procedure with an example, incidentally the same example stated by Hansen. Obtain $x^I_{\min}$ for $A^I x^I \supseteq b^I$, where

$$A^I = \begin{bmatrix} [2, 3] & [0, 1] \\ [1, 2] & [2, 3] \end{bmatrix}, \qquad b^I = \begin{bmatrix} [0, 120] \\ [60, 240] \end{bmatrix}$$

Here, x^1 and x^2 are obtained by solving respectively $A_1 x^1 = \bar{b}$ with $A_1 \in A^I$, and $A_2 x^2 = \underline{b}$ where $A_2 \in A^I$; A_1 and A_2 are suitably chosen, so as to minimize $\|x^1 - x^2\|_E$.

$$A^c = \begin{bmatrix} 5/2 & 1/2 \\ 3/2 & 5/2 \end{bmatrix}, \qquad (A^c)^{-1} = \begin{bmatrix} 5/11 & -1/11 \\ -3/11 & 5/11 \end{bmatrix}$$

The mean values of X_1 are obtained from solving $A^c x = \bar{b}$; for X_2 we have to solve $A^c x = \underline{b}$. This yields $X_1^c = (360/11, 840/11)^T$ and $X_2^c = (-60/11, 300/11)^T$. Now we move from X_1^c and X_2^c in that direction that minimizes the value of $\|x^1 - x^2\|_E$. To obtain x^1, we get the values of the derivatives of all $x^1_{i=1,2}$ with respect to a_{11}, a_{12}, a_{21} and a_{22} to replace into the derivative of $\|x^1 - x^2\|_E^2$ which is

$$D = 2\left(\frac{360}{11} + \frac{60}{11}\right)\frac{\partial x_1^1}{\partial a_{ij}} + 2\left(\frac{840}{11} - \frac{300}{11}\right)\frac{\partial x_2^1}{\partial a_{ij}}$$

whence

$$\frac{\partial x_1^1}{\partial a_{11}} = -[5/11 \quad -1/11]\begin{bmatrix}1 & 0\\ 0 & 0\end{bmatrix}\begin{bmatrix}360/11\\ 840/11\end{bmatrix} = -\frac{1800}{121}$$

$$\frac{\partial x_1^1}{\partial a_{12}} = -[5/11 \quad -1/11]\begin{bmatrix}0 & 1\\ 0 & 0\end{bmatrix}\begin{bmatrix}360/11\\ 840/11\end{bmatrix} = -\frac{4200}{121}$$

$$\frac{\partial x_1^1}{\partial a_{21}} = -[5/11 \quad -1/11]\begin{bmatrix}0 & 0\\ 1 & 0\end{bmatrix}\begin{bmatrix}360/11\\ 840/11\end{bmatrix} = \frac{360}{121}$$

$$\frac{\partial x_1^1}{\partial a_{22}} = -[5/11 \quad -1/11]\begin{bmatrix}0 & 0\\ 0 & 1\end{bmatrix}\begin{bmatrix}360/11\\ 840/11\end{bmatrix} = 840/121$$

$$\frac{\partial x_2^1}{\partial a_{11}} = -[-3/11 \quad 5/11]\begin{bmatrix}1 & 0\\ 0 & 0\end{bmatrix}\begin{bmatrix}360/11\\ 840/11\end{bmatrix} = 1080/121$$

$$\frac{\partial x_2^1}{\partial a_{12}} = -[-3/11 \quad 5/11]\begin{bmatrix}0 & 1\\ 0 & 0\end{bmatrix}\begin{bmatrix}360/11\\ 840/11\end{bmatrix} = 2520/121$$

$$\frac{\partial x_2^1}{\partial a_{21}} = -[-3/11 \quad 5/11]\begin{bmatrix}0 & 0\\ 1 & 0\end{bmatrix}\begin{bmatrix}360/11\\ 840/11\end{bmatrix} = -1800/11$$

$$\frac{\partial x_2^1}{\partial a_{22}} = -[-3/11 \quad 5/11]\begin{bmatrix}0 & 0\\ 0 & 1\end{bmatrix}\begin{bmatrix}360/11\\ 840/11\end{bmatrix} = -4200/121$$

Substituting in the expression for D, we get, after having dropped the common factor 2

$$D_{11} = \frac{420}{11} \times -\frac{1800}{121} + \frac{540}{11} \times \frac{1080}{121} < 0$$

$$D_{12} = \frac{420}{11} \times -\frac{4200}{121} + \frac{540}{11} \times \frac{2520}{121} < 0$$

$$D_{21} = \frac{420}{11} \times \frac{360}{121} + \frac{540}{11} \times -\frac{1800}{121} < 0$$

$$D_{22} = \frac{420}{11} \times \frac{840}{121} + \frac{540}{11} \times -\frac{4200}{121} < 0$$

Whence x^1 is given by

$$\begin{bmatrix}3 & 1\\ 2 & 3\end{bmatrix}\begin{bmatrix}x_1^1\\ x_2^1\end{bmatrix} = \begin{bmatrix}120\\ 240\end{bmatrix}$$

which yields the point (120/7, 480/7). For x^2 we follow the same procedure to obtain the partial derivatives:

$$\frac{\partial x_1^2}{\partial a_{11}} = -[5/11 \quad -1/11]\begin{bmatrix}1 & 0\\0 & 0\end{bmatrix}\begin{bmatrix}-60/11\\300/11\end{bmatrix} = 300/121$$

$$\frac{\partial x_1^2}{\partial a_{12}} = -[5/11 \quad -1/11]\begin{bmatrix}0 & 1\\0 & 0\end{bmatrix}\begin{bmatrix}-60/11\\300/11\end{bmatrix} = -1500/121$$

$$\frac{\partial x_1^2}{\partial a_{21}} = -[5/11 \quad -1/11]\begin{bmatrix}0 & 0\\1 & 0\end{bmatrix}\begin{bmatrix}-60/11\\300/11\end{bmatrix} = -60/121$$

$$\frac{\partial x_1^2}{\partial a_{22}} = -[5/11 \quad -1/11]\begin{bmatrix}0 & 0\\0 & 1\end{bmatrix}\begin{bmatrix}-60/11\\300/11\end{bmatrix} = 300/121$$

$$\frac{\partial x_2^2}{\partial a_{11}} = -[-3/11 \quad 5/11]\begin{bmatrix}1 & 0\\0 & 0\end{bmatrix}\begin{bmatrix}-60/11\\300/11\end{bmatrix} = -180/121$$

$$\frac{\partial x_2^2}{\partial a_{12}} = -[-3/11 \quad 5/11]\begin{bmatrix}0 & 1\\0 & 0\end{bmatrix}\begin{bmatrix}-60/11\\300/11\end{bmatrix} = 900/121$$

$$\frac{\partial x_2^2}{\partial a_{21}} = -[-3/11 \quad 5/11]\begin{bmatrix}0 & 0\\1 & 0\end{bmatrix}\begin{bmatrix}-60/11\\300/11\end{bmatrix} = 300/121$$

$$\frac{\partial x_2^2}{\partial a_{22}} = -[-3/11 \quad 5/11]\begin{bmatrix}0 & 0\\0 & 1\end{bmatrix}\begin{bmatrix}-60/11\\300/11\end{bmatrix} = -1500/121$$

and

$$D_{11} = -\frac{420}{11} \times \frac{300}{121} - \frac{540}{11} \times -\frac{180}{121} < 0$$

$$D_{12} = -\frac{420}{11} \times -\frac{1500}{121} - \frac{540}{11} \times \frac{900}{121} > 0$$

$$D_{21} = -\frac{420}{11} \times -\frac{60}{121} - \frac{540}{11} \times \frac{300}{121} < 0$$

$$D_{22} = -\frac{420}{11} \times \frac{300}{121} - \frac{540}{11} \times -\frac{1500}{121} > 0$$

And x^2 is thence obtained from

$$\begin{bmatrix}3 & 0\\2 & 2\end{bmatrix}\begin{bmatrix}x_1^2\\x_2^2\end{bmatrix} = \begin{bmatrix}0\\60\end{bmatrix}$$

which yields the point (0, 30). Therefore $(x^I)^T = ([0, 120/7], [30, 480/7])$ and

$$A^I x^I = \begin{bmatrix} [2, 3] & [0, 1] \\ [1, 2] & [2, 3] \end{bmatrix} \begin{bmatrix} [0, 120/7] \\ [30, 480/7] \end{bmatrix}$$

$$= \begin{bmatrix} [0, 360/7] + [0, 480/7] \\ [0, 240/7] + [60, 1440/7] \end{bmatrix} = \begin{bmatrix} [0, 120] \\ [60, 240] \end{bmatrix} = b^I$$

the equality to b^I being exact. Thus x^I is the narrowest interval satisfying the relation $A^I x^I \supseteq b^I$. Figure 7 below differentiates the various regions.

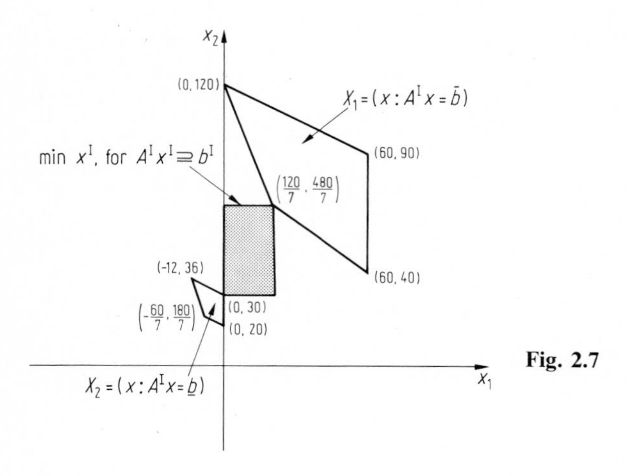

Fig. 2.7

With this, we are through discussing the two problems that we had intended to discuss. Yet, many new problems will still arise in physical applications. Indeed, interval analysis is becoming an invaluable tool in tolerance analysis.

In the foregoing brief discussion, we have restricted ourselves to the cases where A is a square matrix. Naturally, the problem becomes more involved when this matrix A is rectangular. Computation of $(A^I)^+$ will obviously be more complicated. Some work, yet incomplete, has been done on the study of solution structure for linear interval equations when A is rectangular by Ratschek and Sauer (1981). Other special classes of matrices have also attracted some interest; as for example, the class of totally nonnegative interval matrices discussed by Garloff (1980). This class of matrices is encountered in interpolation processes and in problems of approximation. In Rohn (1981 a), the matrices having prescribed column sums are examined; this kind of matrices appears in the input-output description of some economic models.

For their part, numerical analysts only interested in the problem $Ax = b$, can either use Hansen's method to bound the solution inside intervals; thus accounting for round-off, or use an interval method to solve the equations, subsequently using a-posteriori analysis to check the accuracy. If still not satisfied, they may seek better bound-conserving algorithms, or use iterative methods to improve on the solution as we shall be seeing later on.

2.7 An Application to Electrical Networks

In the previous section, we have mentioned a domain within which interval methods can be of great use, namely tolerance analysis. In this section, we shall further exemplify this application with a practical example on filter design in which it is required to find the maximum allowed errors in the designed elements so that the filter's frequency response will not deviate from the ideal one by more than a specified increment.

If $R(x_1, x_2, \ldots, x_n, \omega_i)$ denotes the ideal response of a filter with ideal circuit elements $x_1, x_2, x_3, \ldots, x_n$ at each frequency point ω_i, and if $\tilde{R}(\tilde{x}_1, \tilde{x}_2, \ldots, \tilde{x}_n, \omega_i)$ is the actual response of the filter with circuit elements $\tilde{x}_1, \tilde{x}_2, \ldots, \tilde{x}_n$, then the problem is one of finding the deviation of each component $\tilde{x}_i$ from the ideal x_i, so that $\tilde{R} - R$ will not exceed a certain value $\pm \Delta R(\omega_i)$. It is noteworthy that, had the incremental change $\delta R(\omega_i)$ been precisely known at each frequency ω_i, the problem would have become a simple exercise of fitting $\tilde{R}(\tilde{x}_1, \tilde{x}_2, \ldots, \tilde{x}_n, \omega_i)$ to $R(\omega_i) + \delta R(\omega_i)$, that is

$$\min_{\tilde{x}_1, \ldots, \tilde{x}_n} \sum_{i=1}^{m} [\tilde{R}(\tilde{x}_1, \ldots, \tilde{x}_n, \omega_i) - R(\omega_i) - \delta R(\omega_i)]^2$$

yielding finite values of $\tilde{x}_1, \ldots, \tilde{x}_n$. In other words, there is a one-to-one correspondence between the values of the circuit elements and the particular shape of the response. The unknowns $\tilde{x}_1, \ldots, \tilde{x}_n$ are obtainable using the generalized least-squares method;

$$\tilde{x} = x - (J^T J)^{-1} J^T \delta f$$

where J is some Jacobian matrix and δf is the error between $R(x_1, \ldots, x_n, \omega_i)$ and $R(\omega_i) + \delta R(\omega_i)$. References on the generalized least-squares method include Levenberg (1944), Marquardt (1963) and Fletcher (1969). For the application to filter design, see Deif (1981).

Our problem here is different; $\delta R(\omega_i)$ is unknown, but allowed to vary only between two specified bounds, namely $\pm \Delta R(\omega_i)$. What the range $\tilde{x}_j - x_j$ — i.e. $\pm \Delta x_j$ — is, has yet to be found so that it will verify

$$\Delta R(\omega_i) \geqq \tilde{R}(\tilde{x}_1, \ldots, \tilde{x}_n, \omega_i) - R(x_1, \ldots, x_n, \omega_i) \geqq -\Delta R(\omega_i)$$

For small values of $\pm \Delta R(\omega_i)$, this inequality corresponds to the problem

$$\text{Max } [\Delta x_1], [\Delta x_2], \ldots, [\Delta x_n]$$

subject to the condition that

$$S\,\Delta x \subseteq [-\Delta R(\omega_i), \Delta R(\omega_i)]\,, \qquad i = 1, \ldots, m$$

S being the sensitivity matrix with elements $s_{ij} = \partial R/\partial x_j|_{\omega_i}$ and dimension $m \times n$. This problem was the first to be dealt with in the last section. We will now apply it to the three order Butterworth filter depicted below:

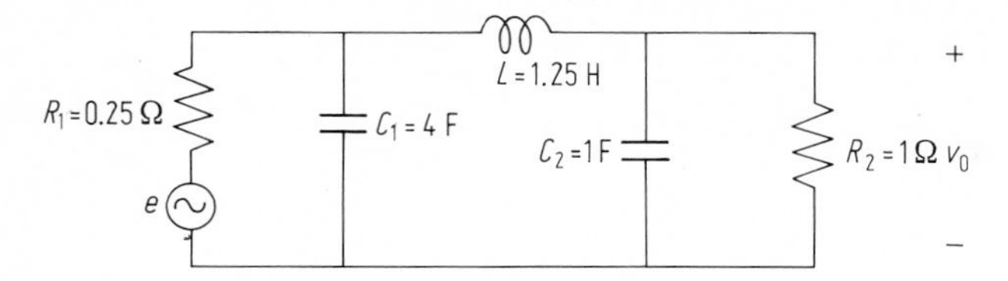

This filter has a frequency response given by

$$R(c_1, L, c_2, \omega) = \left.\frac{v_0(s)}{e(s)}\right|_{s=j\omega}$$

$$= \frac{1}{\sqrt{\left[1 + \frac{R_1}{R_2} - \omega^2\left(\frac{Lc_1R_1}{R_2} + Lc_2\right)\right]^2 + \left[\omega\left(c_1R_1 + c_2R_1 + \frac{L}{R_2}\right) - \omega^3 Lc_1c_2R_1\right]^2}}$$

The value of R has been computed for values of $\omega = 0.2, 0.4, 0.6, 0.8, 1.0, 1.2$, as well as the values of $\partial R/\partial c_1$, $\partial R/\partial L$, $\partial R/\partial c_2$. The problem was to find the ranges of Δc_1, ΔL and Δc_2 that will cause the new response to deviate from the ideal computed by no more than $\pm 1\%$. For this filter, the data were:

$$\begin{bmatrix} -0.004556 & -0.001065 & 0.019403 \\ -0.015826 & -0.018889 & 0.077145 \\ -0.029693 & -0.105347 & 0.145884 \\ -0.044919 & -0.303193 & 0.114969 \\ -0.056569 & -0.452552 & -0.056569 \\ -0.054966 & -0.406901 & -0.171627 \end{bmatrix} \begin{bmatrix} \Delta c_1 \\ \Delta L \\ \Delta c_2 \end{bmatrix} \subseteq \begin{bmatrix} [-0.0079997, & 0.0079997] \\ [-0.0079837, & 0.0079837] \\ [-0.0078197, & 0.0078197] \\ [-0.0071209, & 0.0071209] \\ [-0.0056569, & 0.0056569] \\ [-0.004007, & 0.004007] \end{bmatrix}$$

The matrix on the leftmost side is the sensitivity matrix S evaluated for the three elements at the six frequency points. The vector b on the right-hand side is $\pm\Delta R(\omega_i)$. S^i is calculated to be

$$S^i = \begin{bmatrix} -1\,775.778\,381 & 624.396\,789 & -94.003\,882 & 0 & 0 & 0 \\ -5.981\,765 & 29.262\,743 & -14.678\,855 & 0 & 0 & 0 \\ -365.758\,743 & 148.220\,203 & -22.878\,661 & 0 & 0 & 0 \end{bmatrix}$$

The consistency equations, $SS^ib = b$, come as

$$\left[\begin{array}{ccc|c} & I & & 0 \\ \hline 39.528\,901 & -19.878\,810 & 6.042\,750 & \\ 123.851\,673 & -56.949\,083 & 13.254\,874 & 0 \\ 162.815\,496 & -71.666\,222 & 15.066\,454 & \end{array}\right] b = b$$

Now since the range of values of b_4, b_5, b_6 is narrower than that of b_1, b_2 and b_3, the above consistency condition is inverted and reads

$$\begin{bmatrix} 0.416461 & -0.605201 & 0.365401 \\ 1.323586 & -1.759503 & 1.017087 \\ 1.795386 & -1.829279 & 0.955619 \end{bmatrix} \begin{bmatrix} b_4 \\ b_5 \\ b_6 \end{bmatrix} = \begin{bmatrix} b_1 \\ b_2 \\ b_3 \end{bmatrix}$$

We will now compute the maximum range for Δc_1, found to be given by

$$\begin{aligned} \Delta c_1 &= -1775.778381 b_1 + 624.396789 b_2 - 94.003882 b_3 \\ &= -81.872846 b_4 + 148.034156 b_5 - 103.637235 b_6 \end{aligned}$$

Thus, to maximize Δc_1 is equivalent to minimizing b_4 and b_6 while maximizing b_5. The three variables are then chosen as:

$$\begin{aligned} b_4 &= -0.0071209 \\ b_5 &= 0.0056569 \\ b_6 &= -0.0040070 \end{aligned}$$

These values will unfortunately not satisfy the foregoing consistency conditions; the calculated values of b_1, b_2 and b_3 will be out of range. We still have thence to find out which of b_4, b_5 and b_6 has to be changed to achieve consistency. For this end we check the expressions

$$\frac{\partial \Delta c_1}{\partial b_4} \Bigg/ \left(\frac{\partial b_1}{\partial b_4}, \frac{\partial b_2}{\partial b_4}, \frac{\partial b_3}{\partial b_4}\right)$$

$$\frac{\partial \Delta c_1}{\partial b_5} \Bigg/ \left(\frac{\partial b_1}{\partial b_5}, \frac{\partial b_2}{\partial b_5}, \frac{\partial b_3}{\partial b_5}\right)$$

$$\frac{\partial \Delta c_1}{\partial b_6} \Bigg/ \left(\frac{\partial b_1}{\partial b_6}, \frac{\partial b_2}{\partial b_6}, \frac{\partial b_3}{\partial b_6}\right)$$

The set of minimum values determines which of b_4, b_5 and b_6 will be changed, for the objective function Δc_1 won't change by much, while b_1, b_2 and b_3 change rapidly to become in range. The above quotients are easily calculated by dividing the coefficients of each of b_4, b_5 and b_6 in Δc_1 by its corresponding column in the above consistency matrix, i.e.

Relative sensitivities for	b_4	b_5	b_6
	196.591	−244.6032	283.626
	61.856	84.134	101.896
	45.6018	80.924	108.4503

We find then that b_4 is that which should be altered, followed by b_5 and then b_6. A value of $b_4 = 0.0045672$ is quite sufficient for consistency. Hence,

$$b_4 = 0.0045672, \quad b_5 = 0.0056569, \quad b_6 = -0.0040070$$

and

$$b_1 = -0.0029857, \quad b_2 = -0.0079837, \quad b_3 = -0.0059773$$

Then

$$\Delta c_1 = 0.8788343$$
$$\Delta L = -0.1280253$$
$$\Delta c_2 = 0.0454529$$

Similarly for ΔL,

$$\Delta L = 9.886351 b_4 - 21.01596 b_5 + 13.549591 b_6$$

wherefrom some preliminary values would be

$$b_4 = 0.0071209, \quad b_5 = -0.0056569, \quad b_6 = 0.0040070$$

yielding out-of-range values for b_1, b_2 and b_3. Thence

Relative sensitivities for	b_4	b_5	b_6
	23.738	34.725	37.081
	7.4693	11.944	13.3219
	5.506	11.488	14.1788

Again, we found that it is most suitable to change b_4, which is found to be -0.0045672. Hence

$$b_4 = -0.0045672, \quad b_5 = -0.0056569, \quad b_6 = 0.0040070$$

and

$$b_1 = 0.0029857, \quad b_2 = 0.0079837, \quad b_3 = 0.0059773$$

Then

$$\Delta c_1 = -0.8788343$$
$$\Delta L = 0.1280253$$
$$\Delta c_2 = -0.0454529$$

which is not so surprizing a result, since both ΔL and Δc_1 have opposite sensitivities. As for Δc_2, it is given by

$$\Delta c_2 = 2.78190 b_4 + 2.415306 b_5 - 4.759153 b_6$$

and

$$b_4 = 0.0071209\,, \qquad b_5 = 0.0056569\,, \qquad b_6 = -0.0040070$$

these values need no further alterations, as they satisfy the consistency condition. Then

$$b_1 = -0.0019221\,, \qquad b_2 = -0.0046037\,, \qquad b_3 = -0.0013924$$

yielding

$$\begin{aligned} \Delta c_1 &= 0.6695791 \\ \Delta L &= -0.1027805 \\ \Delta c_2 &= 0.0525198 \end{aligned}$$

Now we are faced with the question as to what range the elements are allowed to vary such that they fulfill the condition

$$-\Delta R \leqq S\,\Delta x \leqq \Delta R$$

The answer to that is any range inside the polyhedron with vertices

$$(\pm 0.8788343,\ \mp 0.1280253,\ \pm 0.0454529)\,,$$
$$(\pm 0.6695791,\ \mp 0.1027805,\ \pm 0.0525198)$$

The remaining vertices are not easy to find. They can of course be evaluated if we allow for different objective functions. This is the technique adopted for linear programming. The results obtained thus far are, however, not to be underestimated. Although the range of errors in the elements cannot be written as $[\pm 0.8788343]$, $[\pm 0.1280253]$ and $[\pm 0.0525198]$ — since this would ensure that $S\,\Delta x^I \supseteq b^I$ as seen in the previous sections — these results still give us a clue as to the maximum range for any one variable. For example, c_1 can be varied until it reaches 4.8788343 as the maximum range allowed for that element, only of course if L and c_2 are also changed to 1.1219747 and 1.0454529 respectively. This result is not achievable through direct observation of S, for Δc_1 can be taken — if $\Delta L = \Delta c_2 = 0$ — as

$$\min_j \Delta b_j / s_{j1}$$

However, this value of Δc_1 will be smaller than that calculated using our method. This is thus a rewarding technique, especially if our selected task was tuning the filter. The element can then be manufactured with a tolerance as large as $\pm 0.88F$

(normalized value to 1 rad/sec cut-off) and the circuit can still be tuned with varying L and c_2 so that the response will not exceed 1%. The reader interested in techniques relevant to this subject can refer to Zereick, Amer and Deif (1983). The authors of the paper were considering finding the minimum number of elements to be tuned starting from some worst case or actual response. The element to be tuned can be selected using the above technique, but its new value must be so chosen that $S\,\Delta x$ will be included in the set X given by $\{\Delta x\colon S\,\Delta x \subseteq [-\Delta R, \Delta R]\}$. This set can only be determined using linear programming techniques.

One last word remaining to be said is that the above technique is only valid for small variations in circuit elements, as only the first order sensitivities are taken into account, neglecting second order terms. It can however be altered for use with large variations by incorporating an interative scheme.

The above discussed method falls under the heading of network design. Application of interval arithmetic to the analysis of electrical networks is an easier task to carry, where one is mainly concerned with the computation of the range of the response function in terms of the variations of the circuit's elements. The reader is referred to Skelboe (1979) for a worst-case analysis of linear electrical circuits.

Exercises 2

1. If A and B are two interval matrices with $A \subseteq B$, show that $\Delta A \leqq \Delta B$ and $|A| \leqq |B|$ where $|A| = \operatorname{Max}\{|\tilde{A}|, \tilde{A} \in A\}$ taken component-wise.

2. Show that $\Delta(A \pm B) = \Delta A + \Delta B$, $\Delta A|B| \leqq \Delta(AB) \leqq \Delta A|B| + |A^c|\,\Delta B$, $|A|\,\Delta B \leqq \Delta(AB) \leqq |A|\,\Delta B + \Delta A|B^c|$.

3. Show how the condition of compatibility of x with the uncertainty in the data of $Ax = b$, namely that $\underline{A}x \leqq \bar{b}$ and $\bar{A}x \geqq \underline{b}$, can be modified for $x_i \leqq 0$.

4. Use one of Hansen's methods to solve $A^I x = b^I$, where

$$A^I = \begin{bmatrix} [1 \pm 0.1] & [-1 \pm 0.1] \\ [1 \pm 0.2] & [-2 \pm 0.3] \end{bmatrix}, \qquad b^I = \begin{bmatrix} [1 \pm 0.2] \\ [-1 \pm 0.1] \end{bmatrix}$$

   Determine the overestimation error.

5. Define the metric $q(A, B) = |A^c - B^c| + |\Delta A - \Delta B|$ measuring the closeness of two interval matrices A and B. Show that

   a) $q(A, B) \geqq 0$
   b) $q(A + C, B + C) = q(A, B)$
   c) $q(A, C) \leqq q(A, B) + q(B, C)$
   d) $q(A + C, B + D) \leqq q(A, B) + q(C, D)$
   e) $q(AC, BC) \leqq q(A, B)\,|C|$
   f) $q(AB, AC) \leqq |A|\,q(B, C)$

6. If $A^I \subseteq \mathbb{R}^{n,n}$ with the matrices $\underline{A}$ and $\bar{A}$ are nonsingular and $\underline{A}^{-1} \geqq 0$ and $\bar{A}^{-1} \geqq 0$, show that $(A^I)^{-1} = [\bar{A}^{-1}, \underline{A}^{-1}]$

7. Show that $w(ab) \leqq w(a)\,\|b\| + w(b)\,\|a\|$, with the bound attained for some a and b. And for $x^I \subseteq (A^I)^{-1}\,b^I$, show how to compare $w(x^I)/\|x^I\|$ with the bound

$$\|A^I\|\,\|(A^I)^{-1}\|\left(\frac{w(A^I)}{\|A^I\|} + \frac{w(b^I)}{\|b^I\|}\right).$$

   For some notions on the conditioning of the problem $A^I x = b^I$, the reader is referred to Zlamal (1977).

8. If $(A^c + \delta A)(x^c + \delta x) = b^c + \delta b$, with $A^c x^c = b^c$ and $|\delta A| \leqq \Delta A$, $|\delta b| \leqq \Delta b$ show that

$$|\delta x| \leqq \frac{|(A^c)^{-1}|}{1 - \| |(A^c)^{-1}| \Delta A \|_1} (\Delta A |x^c| + \Delta b)$$

The reader is to consult Beeck (1975) for further error bounds of interval linear equations.

9. By writing Oettli's inequalities in Sect. 2.4 in the form

$$\sum_j (a^c_{ij} - \Delta a_{ij} \operatorname{sgn}(x_j))\, x_j \leq \bar{b}_i, \qquad i = 1, \ldots, n$$

$$\sum_j (a^c_{ij} - \Delta a_{ij} \operatorname{sgn}(x_j))\, x_j \geq \underline{b}_i, \qquad i = 1, \ldots, n$$

where equalities exist for some n equations out of them, at which some of the x_j's are maximum or minimum, show by choosing all x_j's as the extreme points of the set of feasible solutions that

$$\operatorname{Max} \sum_j (a^c_{ij} - \Delta a_{ij},\, a^c_{ij} + \Delta a_{ij})\, (\underline{x}_j, \bar{x}_j) \geq \bar{b}_i$$

10. If $A = \begin{bmatrix} 2 & 1 \\ 1 & 2 \end{bmatrix}$ and $b = \begin{bmatrix} 1 \\ 3 \end{bmatrix}$

obtain $\hat{x}$ of $Ax = b$ on a three-digit machine with fixed-point arithmetic. Find out whether $\hat{x}$ is compatible or not with the uncertainty in the data. Show how to set the residual r equal to zero by scaling A conveniently. Compare both sets of results.

11. If $|\delta A| \leqq \varepsilon\, |A|$ and $|\delta b| \leqq \varepsilon\, |b|$, where ε is the precision of a calculating machine with floating-point arithmetic used to solve $Ax = b$, show that the error in the solution $\hat{x}$, noted δx, is bounded by $\|\delta x\| \leqq \| |A^{-1}|\, |r| \|$, r being a residual vector equal to $A\hat{x} - b$. Obtain also

$$\frac{\|\delta x\|}{\|x\|} \leqq \varepsilon \frac{\| |A^{-1}|\, |A|\, |\hat{x}| + |A^{-1}|\, |b| \|}{\|x\|}$$

in the case where $\hat{x}$ is compatible with the uncertainty in the data. Compare the results with the bound obtained by Skeel, Sect. 1.5.

12. Check whether the system $Ax = b$, where

$$A = \begin{bmatrix} 10^3 & 2 \times 10^2 & 3 \times 10^2 \\ 5 \times 10^8 & -5 \times 10^9 & 5 \times 10^8 \\ 10^{-7} & 2 \times 10^{-7} & 5 \times 10^{-7} \end{bmatrix}, \qquad b = \begin{bmatrix} 1 \\ 1 \\ 1 \end{bmatrix}$$

is poorly scaled.

13. Define the sets $X_1 = \{x \colon A^I x = b^I\}$, $X_2 = \{x \colon A^I x \subseteq b^I\}$ and $X_3 = \{x \colon A^I \supseteq b^I\}$, where A^I and b^I are given by

$$A^I = \begin{bmatrix} [1, 2] & [-2, 1] \\ [0, 1] & [2, 2] \end{bmatrix}, \qquad b^I = [[-1, 1], [-2, 2]]^T$$

Hence obtain min x^I or max x^I in each.

Chapter 3

Iterative Systems

3.1 Introduction

Whereas direct methods for solving linear equations yield results after a specified amount of computation, iterative methods, in contrast, ameliorate on an approximate solution until it meets a prescribed level of accuracy. In theory, direct methods should give exact solutions in the absence of round-off errors. Yet in practice, due to conditioning problems, this is never the case. For small matrices, direct methods are recommended. For the large ones, direct methods involve very large operations without real need, since the matrices are usually sparse.

Iterative methods have the advantage of simplicity, uniformity and accuracy. They start with an approximation to the solution, and end up, after a few repeated iterations, with a better approximation. They are usually applied to problems in which convergence is known to be rapid or to large scale systems for which direct methods yield inaccurate results, since rounding and truncation occur very frequently in the course of computation. Iterative methods are therefore more adequate for large systems, especially when the system has a sparse matrix. Such systems arise in relation to vibrational problems and in the solution of partial differential equations using finite difference methods.

The only limitation on the use of iterative techniques is that the matrix A should possess enough properties to guarantee convergence; otherwise, we may run into a divergent iteration process. Of such properties, we mention for the time being diagonal-dominance, irreducibility and cyclic qualities; these are not rare in occurrence, which is frequent in practical situations. Apart however from their eventual suitability for the problem in hand, iterative methods are very attractive for numerical analysts in virtue of their great simplicity of use. They only involve matrix addition and multiplication, thus requiring no programming. Furthermore, there is no need for scaling, pivoting, decomposition, elimination or back-substitution, as in direct methods. In short, iterative methods are by far easier to use.

Among the famous iterative methods for the solution of linear equations, we can mention Jacobi's, Gauss-Siedel's and the relaxation types. For each of these methods, the equation $Ax = b$ is rewritten as follows
Jacobi:

$$x^{(i+1)} = (I - D^{-1}A)\, x^{(i)} + D^{-1}b$$

Gauss-Siedel:

$$(D + L)\, x^{(i+1)} = -Ux^{(i)} + b$$

Relaxation:

$$(D + wL)\, x^{(i+1)} = (-wU + (1 - w)\, D)\, x^{(i)} + wb$$

where $A = D + L + U$

$D =$ diagonal matrix of the diagonal entries of A
$L =$ lower matrix containing the elements below the diagonals
$U =$ upper matrix of the elements above the diagonals.

Numerical analysts call w the *relaxation parameter*, and speak of overrelaxation when $w > 1$, and of underrelaxation when $w < 1$. For $w = 1$, the Gauss-Siedel is recovered.

Therefore, starting from an initial vector $x^{(0)}$, one generates a sequence of vectors $x^{(0)}, x^{(1)}, x^{(2)}, \ldots$ which converge towards the desired solution x. The proof of convergence is straightforward; for instance for the Jacobi method, writing the recurrence relation in the form $x^{(k)} = Jx^{(k-1)} + h$ with J and h defined as in above, implies in conjunction with $x = Jx + h$, that

$$x^{(k)} - x = J^k(x^{(0)} - x)$$

showing that $x^{(k)}$ will approach x for every $x^{(0)}$ if and only if $\varrho(J) < 1$. An alternative sufficient condition could be expressed as $\|J\| < 1$, which under the various norms used imposes different conditions on the elements of A (c.f. Faddeeva (1959)). In fact, in real practical problems, we find that the conditions of convergence are guaranteed. For example, in boundary value problems treated using the finite difference approximation, A is *diagonal-dominant*, that is

$$|a_{ii}| > \sum_{\substack{j=1\\ j\neq i}}^{n} |a_{ij}| \quad \text{or} \quad |a_{jj}| > \sum_{\substack{i=1\\ i\neq j}}^{n} |a_{ij}|$$

This ensures that $\|J\| < 1$ for Jacobi, relative to an l_∞ or l_1-norm (see exercise 3.7), and that $\varrho(G = -(D + L)^{-1}\, U) < 1$ for Gauss-Siedel (see exercise 3.4). Even if A is not diagonal-dominant, the above methods still converge for some lesser restrictions. For instance, for the class of positive-definite matrices, $\varrho(G) < 1$ (see exercise 3.8). Also, if A is irreducible, we can be satisfied with a weak row or column criterion, and so on ...

In this chapter, we will neither survey nor compare the different iterative methods. For such a treatment of the subject, the reader may refer to the excellent treatises by Varga (1962). Householder (1964), Young (1971) and Young and Gregory (1973). Our efforts will be directed to performing sensitivity analysis of the equations. The reason for assigning a separate chapter for this category of methods, apart from its being a different way of treatment, is that they involve two kinds of errors, one of which being specific to iterations. Indeed, while the first type of errors is due to perturbations in the coefficients, as previously seen for direct methods, the second type relates to the choice of a stoppage criterion to terminate the iteration process; the iterations as we know are only finite in number. On the other hand, as we will be finding, the rates of convergence of the iterative methods which depend on $\varrho(J)$, are also a function of the conditioning of the problem. Apart from worsening results because of round-off, the condition number also reduces the rates of convergence. A method to reduce cond (A) based on order reduction for large systems will also be discussed. It is based on keeping only those equations which are well conditioned. At last, we will consider and discuss an iterative system in which the matrix A is an interval matrix.

3.2 An Alternative Bound

Supposing we wish to perform an error analysis of the equation

$$x = D^{-1}b + (I - D^{-1}A)\,x$$

it is obvious that we should obtain the same bounds as those derived in section 1.4, whatever the shape of the equations. Now since, for some norm,

$$\|I - D^{-1}A\| < 1$$

an error bound incorporating such a quantity would clarify more the sensitivity of x to changes in A and b. In many practical situations, A is strongly diagonal dominant, that is

$$\|I - D^{-1}A\| \ll 1$$

and the Jacobi iterations become very stable and even very fast. To ascertain the first proposition, we will follow the same elementary routine as before, and from $x = D^{-1}b + (I - D^{-1}A)\,x$, obtain

$$\Delta x = D^{-1}\,\Delta b - D^{-1}\,\Delta D D^{-1}b + (I - D^{-1}A)\,\Delta x + D^{-1}\,\Delta D D^{-1}Ax - D^{-1}\,\Delta A x$$

to a first order approximation. By cancelling redundant terms, we get

$$\Delta x = D^{-1}\,\Delta b + (I - D^{-1}A)\,\Delta x - D^{-1}\,\Delta A x$$

Hence

$$\|\Delta x\|\,(1 - \|I - D^{-1}A\|) \leqq \|D^{-1}\|\,\|\Delta b\| + \|D^{-1}\|\,\|\Delta A\|\,\|x\|$$

or alternatively

$$\frac{\|\Delta x\|}{\|x\|} \leq \frac{1}{1 - \|I - D^{-1}A\|}\left(\|D^{-1}\|\,\|\Delta A\| + \frac{\|D^{-1}\|\,\|\Delta b\|}{\|x\|}\right)$$

But given that

$$\|x\|\,(1+\|I-D^{-1}A\|) \geqq \|x-(I-D^{-1}A)\,x\| = \|D^{-1}b\| \geqq \|D\|^{-1}\,\|b\|$$

we have therefore that

$$\frac{\|\Delta x\|}{\|x\|} \leq \frac{1+\|I-D^{-1}A\|}{1-\|I-D^{-1}A\|}\left(\frac{\|\Delta A\|\,\|D^{-1}\|}{1+\|I-D^{-1}A\|}+\frac{\|D\|\,\|D^{-1}\|\,\|\Delta b\|}{\|b\|}\right).$$

Here, we note that

$$1+\|I-D^{-1}A\| \geqq \|D^{-1}A\| \geqq \|D\|^{-1}\,\|A\|$$

And by putting

$$\|D\|\,\|D^{-1}\| = \text{cond}\,(D) = \max_i |a_{ii}|/\min_i |a_{ii}|$$

we finally obtain

$$\frac{\|\Delta x\|}{\|x\|} \leq \left(\frac{1+\|I-D^{-1}A\|}{1-\|I-D^{-1}A\|}\right)\left(\frac{\max_i |a_{ii}|}{\min_i |a_{ii}|}\right)\left(\frac{\|\Delta A\|}{\|A\|}+\frac{\|\Delta b\|}{\|b\|}\right)$$

The result is self-explanatory. Note that the product of the first two brackets on the right-hand side of the inequality sign must be of the same order as cond (A), and in fact they are, for, when writing A in the form $A = D - D(I - D^{-1}A)$, we obtain

$$\|A\| \leqq \|D\|\,(1+\|I-D^{-1}A\|)$$

and

$$\|A^{-1}\| \leqq \frac{\|D^{-1}\|}{1-\|I-D^{-1}A\|}$$

as set forth in exercise 1.18. From this, it follows that

$$\begin{aligned}\text{cond}\,(A) &= \|A\|\,\|A^{-1}\| \\ &\leq \|D\|\,\|D^{-1}\|\left(\frac{1+\|I-D^{-1}A\|}{1-\|I-D^{-1}A\|}\right) \\ &= \frac{\max_i |a_{ii}|}{\min_i |a_{ii}|}\left(\frac{1+\|I-D^{-1}A\|}{1-\|I-D^{-1}A\|}\right).\end{aligned}$$

(cf. Varah (1975), for different norms).

We can therefore conclude that the numerical calculations yield more accurate results whenever the diagonal elements are of comparable size, with $\|I - D^{-1}A\|$ much smaller than unity, i.e. the sum of the off-diagonal elements in absolute value is very small compared to the diagonal ones or, in short, with A being strongly diagonal dominant. Such matrices yield small errors in the solution and cond (A) will thus approach unity, which is its lowest attainable value.

Indeed, iterative methods guarantee the existence of the above qualities, for writing the equations $Ax = b$ in the form $x = D^{-1}b + (I - D^{-1}A)\,x$ is in fact synonymous to scaling them. The latter form, when read $x = h + Jx$, sets $I - J$ as the new coefficients' matrix, and h as the new right-hand side. And because $I - J$ is a diagonal dominant matrix with unity diagonal elements, the solution becomes less sensitive to perturbations in either J or h. Assuming enough iterations are performed, the relative error in the solution will not exceed the following bound

$$\frac{\|\Delta x\|}{\|x\|} \leq \frac{1 + \|J\|}{1 - \|J\|}\left(\frac{\|\Delta J\|}{1 + \|J\|} + \frac{\|\Delta h\|}{\|h\|}\right)$$

For example, given that

$$A = \begin{bmatrix} 10^3 & 2 \times 10^2 & 3 \times 10^2 \\ 5 \times 10^8 & -5 \times 10^9 & 5 \times 10^8 \\ 10^{-7} & 2 \times 10^{-7} & 5 \times 10^{-7} \end{bmatrix}, \qquad b = \begin{bmatrix} 22 \times 10^3 \\ 345 \times 10^9 \\ 29 \times 10^{-6} \end{bmatrix}$$

cond $(A) \cong 10^{16}$; the solution on a ten-digit machine with floating-point arithmetic is given by

$$\hat{x} = (10.00000003, -60.00000000, 79.99999997)^T$$

when we use a direct elimination method. The residual $r = A\hat{x} - b$ was found to be

$$r = (2 \times 10^{-5}, 0, -10^{-14})^T$$

For the same system, when processed in the iterative form, we get

$$\begin{bmatrix} x_1 \\ x_2 \\ x_3 \end{bmatrix} = \begin{bmatrix} 0 & -0.2 & -0.3 \\ 0.1 & 0 & 0.1 \\ -0.2 & -0.4 & 0 \end{bmatrix} \begin{bmatrix} x_1 \\ x_2 \\ x_3 \end{bmatrix} + \begin{bmatrix} 22 \\ -69 \\ 58 \end{bmatrix}$$

Cond $(I - J)$ is then given by

$$\text{cond}\,(I - J) \leqq \frac{1 + \|J\|}{1 - \|J\|} = \frac{1.6}{0.4} = 4$$

After 20 iterations, the solution is given by

$$\hat{x} = (10, -60, 80)^T$$

which is the exact value for the solution. Even for a smaller number of iterations, say $k = 17$; the solution is

$$\hat{x} = (10.00000001, -60.0, 80)^T$$

with a residual r given by

$$\begin{aligned} r &= \hat{x} - J\hat{x} - h \\ &= (10^{-8}, -10^{-9}, 2 \times 10^{-9})^T \end{aligned}$$

Supposing now we didn't know the exact solution beforehand, which is usually the case, and we wish to compare both results obtained from calculations, then the comparison of the two quantities

$$\|A^{-1}\| \, \|A\| \frac{\|r\|}{\|b\|} \quad \text{for the first system}$$

$$\frac{1 + \|J\|}{1 - \|J\|} \frac{\|r\|}{\|h\|} \quad \text{for the second system}$$

would reveal that the second system's output is more accurate. In fact, for the second system,

$$\frac{\|\Delta x\|}{\|x\|} \leq \frac{1 + \|J\|}{1 - \|J\|} \cdot \frac{\|r\|}{\|h\|} = 4\left(\frac{10^{-8}}{69}\right) \simeq 5.8 \times 10^{-10}$$

And since the machine's precision reaches at best 5×10^{-10}, we can envision how accurate the solution obtained is. This comes not as a surprise since the first system has a large valued cond (A); alteration of only the tenth decimal digit produces a larger residual. This becomes very clear when we substitute the solution of the second system into the first. The residual obtained is

$$r = (10^{-5}, 5, 10^{-15})^T$$

Still now, we are faced with a perplexing situation. The computed residual of the first system, corresponding to the first solution, is smaller than that of the same system that corresponds to the second solution: $r_1 = (2 \times 10^{-5}, 0, -10^{-14})^T$ and $r_2 = (10^{-5}, 5, 10^{-15})^T$. Does this mean that, for the first system, the first solution is more accurate than the second one? And even this phrasing is still incorrect, for, as we have seen in Sect. 2.5, a smaller residual (within a certain machine accuracy) does not necessarily imply a better solution. A correct comparison is established using the a-posteriori error bound given by

$$\|\Delta x\| \leqq \| \, |A^{-1}| \, |r| \, \|$$

(see also exercise 3.18) and as A^{-1} is given by

$$A^{-1} = \begin{bmatrix} 0.001\,056\,911 & 1.626\,016\,239 \times 10^{-11} & -650\,406.5040 \\ 0.000\,081\,301 & -1.910\,569\,106 \times 10^{-10} & 142\,276.4228 \\ -0.000\,243\,902 & 7.317\,073\,161 \times 10^{-11} & 2\,073\,170.732 \end{bmatrix}$$

we get

$$\| |A^{-1}|\,|r_1| \| \cong 2.8 \times 10^{-8}$$
$$\| |A^{-1}|\,|r_2| \| \cong 1.1 \times 10^{-8}$$

The above result is based on the l_∞-norm. This means that the second solution obtained by iteration is more accurate, even for the first system. In all cases, although the second solution is better, the first one still is acceptable. The true error in the solution by direct elimination is

$$\frac{\|\Delta x\|}{\|x\|} = 3 \times 10^{-8}/80 = 3.75 \times 10^{-10}$$

based on an l_∞-norm. This is a very good result, the error being small. One may wonder why the solution of the first system obtained by elimination is still so good, although cond $(A) \cong 10^{16}$. By considering the realistic definition of cond (A) as cond $(A) = \| |A^{-1}|\ |A| \|$ (cf. Sect. 1.5), which yields a value of 2.46 for the system, this fact can be explained, as the first system is well-conditioned. Strongly diagonal dominant matrices are well-conditioned irrespective of any possible disparity in the size of the diagonal elements, given that

$$\begin{aligned} \| |A^{-1}|\,|A| \| &= \| |(I - J)^{-1} D^{-1}|\,|D(I - J)| \| \\ &= \| |(I - J)^{-1}|\,|I - J| \| \end{aligned}$$

This means that both direct and iterative systems have a comparable condition number. The error in the solution is governed by the bound (see Sect. 1.5)

$$\begin{aligned} \|\Delta x\| &\leqq \| |(I - J)^{-1}|\,|r| \| \\ &\leqq \| |(I - J)^{-1}|\,|\Delta J|\,|\hat{x}| + |(I - J)^{-1}|\,|\Delta h| \| \end{aligned}$$

3.3 Rates of Convergence

In the previous section, we have demonstrated that diagonal dominant matrices with small-valued $\|I - D^{-1}A\|$ yield accurate solutions. This latter quantity does not only affect the accuracy of the result, but also the rate at which the solution converges.

From the recurrence relation $x^{(k)} = Jx^{(k-1)} + h$, we can calculate the error vector at the kth iteration as

$$x^{(k)} - x = J^k(x^{(0)} - x)$$

The rate of convergence of the term $x^{(k)}$ towards the solution x depends thus on J^k. In fact, by taking the norms, i.e.

$$\|x^{(k)} - x\| \leqq \|J^k\| \, \|x^{(0)} - x\|$$

we can conclude that the *average reduction factor per iteration*, namely $(\|x^{(k)} - x\|/\|x^{(0)} - x\|)^{1/k}$, written for the successive error norms, is bounded by the norm $\|J^k\|^{1/k}$. Either of the two quantities can be taken as a measure for the rate of convergence. For example, by writing

$$\left(\frac{\|x^{(k)} - x\|}{\|x^{(0)} - x\|}\right)^{1/k} = e^{-\left\{-\frac{1}{k}(\ln(\|x^{(k)} - x\|/\|x^{(0)} - x\|))\right\}}$$

the *average rate of convergence for k iterations* becomes

$$R = -\frac{1}{k}(\ln(\|x^{(k)} - x\|/\|x^{(0)} - x\|))$$

(cf. Kahan (1958)). A more representative average rate is given by

$$R = -\frac{1}{k}\ln \|J^k\|$$

It allows one to compare any two distinct systems of different matrix J. For example, a system $x = J_1 x + f_1$ would have a higher rate of convergence than another, say $x = J_2 x + f_2$, if $\|J_1^k\| < \|J_2^k\|$ for a fixed number of iterations k.

The above rate of convergence is a function of the number of iterations k at the outset of which it is evaluated. Instead, Young (1954) proposed $-\ln \varrho$ as the *asymptotic rate of convergence*, where ϱ is the spectral radius of J. In fact,

$$\lim_{k \to \infty} \|J^k\|^{1/k} = \varrho$$

This follows immediately from the relation

$$J^k = T \left[\begin{array}{cccc|ccc} \lambda_1^k & k\lambda_1^{k-1} & \ldots & \binom{k}{m-1} \lambda_1^{k-m+1} & & & \\ & \lambda_1^k & \ddots & & & 0 & \\ & & & \lambda_1^k & & & \\ \hline & & & & \lambda_2^k & & \\ & & 0 & & & \ddots & \\ & & & & & & \lambda_n^k \end{array}\right] T^{-1}, \qquad |\lambda_1| = \varrho$$

(with the assumption that λ_1, in the general case, belongs to a nonlinear elementary divisor of degree m) since by taking norms, we get

$$\binom{k}{m-1} \varrho^{k-m+1}/\|T\| \, \|T^{-1}\| \leq \|J^k\| \leq \binom{k}{m-1} \varrho^{k-m+1} \|T\| \, \|T^{-1}\|$$

for large values of k. And for $k \to \infty$, we have that

$$\binom{k}{m-1}^{1/k} \to 1$$

$$\|T\|^{1/k} \, \|T^{-1}\|^{1/k} \to 1$$

and therefore

$$\|J^k\|^{1/k} \to \varrho$$

$\|T\| \, \|T^{-1}\|$ is the condition number of the eigenvectors of J; it enters into the expressions for error bounds in eigenvalue problems. However, by keeping the index k, the average rate of convergence after k iterations is given by

$$-\frac{1}{k} \ln \|J^k\| = -\frac{1}{k}\left\{\ln \binom{k}{m-1} + \ln \varrho^{k-m+1} + \ln v\right\}$$

with

$$\frac{1}{\text{cond}\,(T)} \leqq v \leqq \text{cond}\,(T)$$

However, if J is normal ($J^*J = JJ^*$), then J only possesses linear divisors, and the first term of the right-hand side vanishes. The second term becomes only $\ln \varrho^k$ while the third reduces to zero (T is unitary, meaning that cond $(T) = 1$ relative to l_2-norm). The average rate of convergence is thus independent of k and equals always $-\ln \varrho$ irrespective of the value k assumes.

Now, what effect does cond $(I - J)$ exert on the rate of convergence? We can state that the greater cond $(I - J)$, the smaller is the rate of convergence, for we have simply that

$$\text{cond}\,(I - J) \leqq \frac{1 + \|J\|}{1 - \|J\|}$$

Thenceforth, the greater cond $(I - J)$, the greater is $\|J\|$. Nevertheless, this does not guarantee a large value $\|J^k\|^{1/k}$; it only serves as a crude estimate of the smallest rate of convergence, since $-\ln \varrho \geqq -\ln \|J\|$ ($\varrho \leqq \|J\|$). A better estimate is based on ϱ itself. In fact, for special classes of matrices, Gauss-Siedel's $\varrho(G = -(D + L)^{-1} U)$ and the relaxation method's $\varrho(S = (D + wL)^{-1} (-wU + (1 - w)D))$ both depend on $\varrho(J)$. We should therefore conclude that the slower the Jacobi method is, the slower the other methods are. For this relation, we define matrices with *property A* according to Young (1950) or its generalization, owed to Varga (1962), which he termed: *the class of consistently ordered matrices.*

According to Young, A is termed of property A if there exists a permutation matrix P such that

$$PAP^T = \left[\begin{array}{c|c} D_1 & M_1 \\ \hline M_2 & D_2 \end{array}\right]$$

D_1 and D_2 being diagonal matrices. Such matrices possess interesting properties. If A is of property A, then $\bar{A} = PAP^T$ is consistently ordered, i.e. the matrix $\bar{J}$ of Jacobi's, obtained from $\bar{A}$ as

$$\bar{J} = I - \bar{D}^{-1}\bar{A} = -\bar{D}^{-1}(\bar{L} + \bar{U}) = \left[\begin{array}{c|c} 0 & -D_1^{-1}M_1 \\ \hline -D_2^{-1}M_2 & 0 \end{array}\right]$$

can still be written in the form

$$\bar{J}(\alpha) = -\bar{D}^{-1}\left(\alpha\bar{L} + \frac{1}{\alpha}\bar{U}\right)$$

with eigenvalues that are independent of α. The reason is that

$$\bar{J}(\alpha) = \left[\begin{array}{c|c} 0 & -\frac{1}{\alpha}D_1^{-1}M_1 \\ \hline -\alpha D_2^{-1}M_2 & 0 \end{array}\right]$$

$$= \left[\begin{array}{c|c} I & 0 \\ \hline 0 & \alpha I_2 \end{array}\right]\left[\begin{array}{c|c} 0 & -D_1^{-1}M_1 \\ \hline -D_2^{-1}M_2 & 0 \end{array}\right]\left[\begin{array}{c|c} I_1 & 0 \\ \hline 0 & \alpha I_2 \end{array}\right]^{-1}$$

i. e. $\bar{J}(\alpha)$ and $\bar{J}(1)$ are similar, thus having the same eigenvalues. Note that a consistently ordered matrix need not be written in the above $\bar{A}$ form. For example, the tridiagonal matrix

$$\begin{bmatrix} 1 & a & 0 \\ b & 1 & c \\ 0 & d & 1 \end{bmatrix}$$

is consistently ordered. Also, if A is of property A, this doesn't imply that A is consistently ordered, or vice versa. For example

$$A = \begin{bmatrix} 1 & -1/4 & 0 & -1/4 \\ -1/4 & 1 & -1/4 & 0 \\ 0 & -1/4 & 1 & -1/4 \\ -1/4 & 0 & -1/4 & 1 \end{bmatrix}$$

is of property A, but it is inconsistently ordered. Rather, if A is of property A, then it is consistently ordered if and only if

$$P\{D(I - J(\alpha))\}\, P^T = P\left\{D\left(I + D^{-1}\left(\alpha L + \frac{1}{\alpha}U\right)\right)\right\} P^T = \left[\begin{array}{c|c} D_1 & \frac{1}{\alpha}M_1 \\ \hline \alpha M_2 & D_2 \end{array}\right]$$

i.e.

$$PLP^T = \begin{bmatrix} 0 & 0 \\ M_2 & 0 \end{bmatrix}, \qquad PUP^T = \left[\begin{array}{c|c} 0 & M_1 \\ \hline 0 & 0 \end{array}\right]$$

Hence, if A is of property A as well as consistently ordered, then

$$PJ(\alpha)\,P^T = \left[\begin{array}{c|c} 0 & -\dfrac{1}{\alpha}D_1^{-1}M_1 \\ \hline -\alpha D_2^{-1}M_2 & 0 \end{array}\right]$$

for all values of α. And apart from $J(\alpha)$ being possibly permuted into the right-hand side's form, in the above expression, its eigenvalues are independent of α.

One very important property of the class of consistently ordered matrices: if μ is an eigenvalue of $J(\alpha)$, then so will be $-\mu$. This is proved using the fact that μ is the root of

$$\left|\mu I + D^{-1}\left(\alpha L + \frac{1}{\alpha}U\right)\right| = 0$$

and then by taking $\alpha = 1, -1$. This means that if A is consistently ordered, then the eigenvalues of J are always $\pm\mu_i$, these latters being also the same eigenvalues of $J(\alpha)$, whatever the value α may assume. However, if A is of property A, then J has $\pm\mu_i$ for eigenvalues, whereas $J(\alpha)$ does not necessarily for all values of α. In such cases, the above congruent transformation of $J(\alpha)$ is only valid for $\alpha = \pm 1$.

Let us consider now the relation between the three methods, namely Jacobi's, Gauss-Siedel's and the relaxation's in terms of their respective spectral radii. We shall assume that A is generally of property A also being consistently ordered. We shall also take J to be normal to facilitate our results in terms of $\varrho(J)$. The results will be ultimately formulated in terms of spectral norm. We have that

1. Jacobi:

$$\operatorname{cond}(I - J) = \frac{1 + \varrho(J)}{1 - \varrho(J)}$$

2. Gauss-Siedel:

$$PJP^T = \begin{pmatrix} 0 & F_1 \\ F_2 & 0 \end{pmatrix}$$

wherein, for normal J, $F_1^*F_1 = F_2F_2^*$ and $F_2^*F_2 = F_1F_1^*$. Taking $G = -(D + L)^{-1}U$ as our coefficients matrix, we get

$$PGP^T = \begin{bmatrix} 0 & F_1 \\ 0 & F_2F_1 \end{bmatrix}$$

and

$$PG^kP^T = \begin{bmatrix} 0 & F_1(F_2F_1)^{k-1} \\ 0 & (F_2F_1)^k \end{bmatrix}$$

whence

$$\|G^k\| = \left\| \begin{matrix} 0 & F_1(F_2F_1)^{k-1} \\ 0 & (F_2F_1)^k \end{matrix} \right\| = \left\| \begin{bmatrix} 0 & F_1 \\ 0 & F_2F_1 \end{bmatrix} \begin{bmatrix} 0 & 0 \\ 0 & (F_2F_1)^{k-1} \end{bmatrix} \right\|$$

$$= \text{root square } \hat{\lambda} \left(\begin{bmatrix} 0 & 0 \\ 0 & (F_2F_1)^{*k-1} \end{bmatrix} \begin{bmatrix} 0 & 0 \\ F_1^* & (F_2F_1)^* \end{bmatrix} \right.$$

$$\left. \times \begin{bmatrix} 0 & F_1 \\ 0 & F_2F_1 \end{bmatrix} \begin{bmatrix} 0 & 0 \\ 0 & (F_2F_1)^{k-1} \end{bmatrix} \right)$$

But as the matrices commute, J being normal, then

$$\|G^k\| = \sqrt{\varrho^{k-1}(F_1^*F_2^*F_2F_1)}\ \sqrt{\varrho(F_1^*F_1 + F_1^*F_2^*F_2F_1)}$$

$$= \sqrt{\varrho^{2k-2}(F_1^*F_1)}\ \sqrt{\varrho(F_1^*F_1)}\ \sqrt{1 + \varrho(F_1^*F_1)}$$

And as $\varrho(J) = \sqrt{\varrho(F_1^*F_1)}$, we finally obtain

$$\|G^k\| = \varrho^{2k-1}(J)\sqrt{1 + \varrho^2(J)}$$

This result was obtained by Wozniakowski (1978) for symmetric J. Here it is generalized for the class of normal matrices J. Note also that $\|G^k\|^{1/k} \to \varrho^2(J)$, implying that Gauss-Siedel's method is twice as fast in converging as the Jacobi method.

3. Successive overrelaxation (S.O.R.)

$$S = (D + wL)^{-1}(-wU + (1 - w)D)$$

It can easily be shown, as in Stoer and Bulirsch (1980), that when A is consistently ordered, we have

$$(\lambda + w - 1)^2 = \lambda w^2 \mu^2$$

where λ is an eigenvalue of S and μ is an eigenvalue of J. The above result is due to Young (1950). A generalization to the class of p-cyclic matrices can be found in Varga (1962).

The question arises now as to which value of w would ensure that $\varrho(S)$ is minimal. Young (1950) obtained the optimal solution from the above relation, viz.

$$w_b = \frac{2}{1 + \sqrt{1 - \varrho^2(J)}}$$

As for the optimal value of $\varrho(S)$, noted $\varrho_b(S)$, it is given by

$$\varrho_b(\text{SOR}) = w_b - 1 = \frac{1 - \sqrt{1 - \varrho^2(J)}}{1 + \sqrt{1 + \varrho^2(J)}}$$

The reason for the notation $\varrho(SOR)$, SOR being the abbreviation for successive overrelaxation, is that $w_b > 1$. In fact, Kahan showed much earlier, in 1958, that $\varrho(S) \geqq |w - 1|$ whatever the value of w, i.e. that values of w with $0 < w < 2$, at best, lead to convergent methods. This narrowed considerably the space for other later authors to search for w_b. By substituting for $\varrho(J) = \text{cond}\,(I - J) - 1/\text{cond}\,(I - J) + 1$, we get

$$\varrho_b(\text{S.O.R.}) = \left(\frac{\sqrt{\text{cond}\,(I - J)} - 1}{\sqrt{\text{cond}\,(I - J)} + 1}\right)^2$$

This is very similar a result to that obtained by Wozniakowski (1978), for J normal. Apart from its worsening the result due to the large value of cond $(I - J)$ appearing, it looks as though this method, like the two previous ones, will have a small rate of convergence.

3.4 Accuracy of Solutions

In Sect. 3.2, we have seen that the relative error in the solution x, noted Δx, is bounded by

$$\frac{\|\Delta x\|}{\|x\|} \leqq \text{cond}\,(I - J) \cdot \varepsilon$$

where ε is the machine precision with respect to $\|J\|$. Unfortunately the error in x as such is usually complemented by an additional one resulting from the fact that the number of iterations k is finite. Therefore, we end up with two kinds of errors: those due to uncertainties in the data, and those due to interruption of the iterations. To obtain a realistic error bound should, in the light of this fact, mean to obtain a compound error bound accounting for both types.

The following figure will be used to clarify this notion of error.

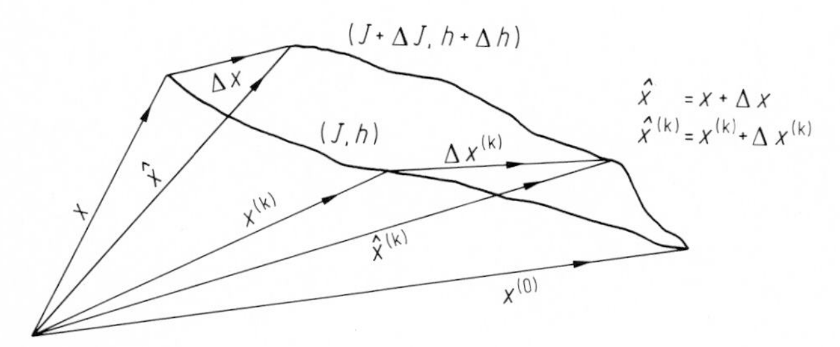

Fig. 3.1

Due to the fact that ΔJ and Δh are not null, the actual solution $\hat{x}^{(k)}$ will differ after each iteration k from $x^{(k)}$ by an amount $\Delta x^{(k)}$ which depends on ΔJ and Δh. The relation between those three variables, namely $\Delta x^{(k)}$, ΔJ and Δh, can be deduced from the relations

$$x^{(k)} + \Delta x^{(k)} = (J + \Delta J)\,(x^{(k-1)} + \Delta x^{(k-1)}) + h + \Delta h$$

and

$$x = Jx + h\,.$$

We obtain

$$\begin{aligned} x^{(k)} + \Delta x^{(k)} - x &= (J + \Delta J)\,(x^{(k-1)} + \Delta x^{(k-1)} - x) + \Delta J x + \Delta h \\ &= (J + \Delta J)^k\,(x^{(0)} - x) + \sum_{i=0}^{k-1} (J + \Delta J)^i\,(\Delta J x + \Delta h) \end{aligned}$$

wherefrom it follows that

$$\Delta x^{(k)} = (J + \Delta J)^k\,(x^{(0)} - x) - J^k(x^{(0)} - x) + \sum_{i=0}^{k-1} (J + \Delta J)^i\,(\Delta J x + \Delta h)$$

Here, one should note that for $k \to \infty$,

$$\Delta x = (I - [J + \Delta J])^{-1}\,(\Delta J x + \Delta h)$$

as expected. As for the total error after any iteration k, we obtain

$$\hat{x}^{(k)} - x = (J + \Delta J)^k\,(x^{(0)} - x) + (I - [J + \Delta J])^{-1}\,(I - J^k)\,(\Delta J x + \Delta h)$$

And, assuming $x^{(0)} = 0$ for unification, we finally get

$$\frac{\|\hat{x}^{(k)} - x\|}{\|x\|} \leq \|J\|^k \left(1 + \frac{\|\Delta J\|}{\|J\|}\right)^k + \frac{1 + \|J\|}{1 - \|J\| - \|\Delta J\|} \left(\frac{\|\Delta J\|}{1 + \|J\|} + \frac{\|\Delta h\|}{\|h\|}\right) (1 + \|J\|^k)$$

From which the bound in Sect. 3.2 follows, as a special case occurring when $k \to \infty$.

In practice, what happens is that for large values of k, due to the fixed word length on the display, the error due to stoppage of the iteration process becomes negligible relative to that due to rounding. The effect of truncating the iterations becomes only pronounced during the early stages of the computation.

Still, workers have been more interested in $\hat{x}^{(k)}$ than in $\hat{x}$, since the former is involved in the evaluation of the residual vector after k iterations, which is given by

$$r^{(k)} = \hat{x}^{(k)} - J\hat{x}^{(k)} - h$$

From this relation and that stating that

$$0 = x - Jx - h$$

we get the following a-posteriori bound, valid after a number of iterations k,

$$\frac{\|\hat{x}^{(k)} - x\|}{\|x\|} \leq \frac{1 + \|J\|}{1 - \|J\|} \cdot \frac{\|r^{(k)}\|}{\|h\|}$$

which is similar to that obtained in Sect. 3.2. This bound computes the accuracy of $\hat{x}^{(k)}$ relative to the exact solution x once $r^{(k)}$ is calculated at the final iteration k.

The above bound can also be written in terms of the eigenvalues of J in the form

$$\|\hat{x}^{(k)} - x\| \leq \|(I - J)^{-1}\| \, \|r^{(k)}\| \leq \|r^{(k)}\| \cdot v \cdot \max_i \left| \frac{1}{1 - \lambda_i} \right|; \qquad v \geq 1$$

with $v = 1$ for J normal, relative to an l_2-norm. And because the stopping criterion is usually imposed on $\hat{x}^{(k+1)} - \hat{x}^{(k)}$ rather than $r^{(k)}$, both quantities being in fact equal if we neglect rounding, then for $\|\hat{x}^{(k+1)} - \hat{x}^{(k)}\| < \alpha$, we have

$$\|\hat{x}^{(k)} - x\| \leq v\alpha \max_i \left| \frac{1}{1 - \lambda_i} \right| + 0(\|\Delta J\|, \|\Delta h\|)$$

This means that if the iterations are terminated when $\|\hat{x}^{(k+1)} - \hat{x}^{(k)}\| < \alpha$ and $\|\hat{x}^{(i+1)} - \hat{x}^{(i)}\| \geqq \alpha (i < k)$ for some assigned $\alpha > 0$, then the above inequality sets a simple bound on how small the length of the final error vector $\hat{x}^{(k)} - x$ will be, i.e. on the difference between the obtained solution and the exact one. The reader interested in knowing more about error distribution in relation to stoppage criteria is referred to Yamamoto (1975, 1976).

Supposing even that $r^{(k)}$ is restricted to some criterion, how can one make sure that the value of $\hat{x}^{(k)}$ computed is admissible with respect to the uncertainties in the data ΔJ and Δh? In other words, what is the maximum allowable value for $|r^{(k)}|$ at the kth iteration for $\hat{x}^{(k)}$ to be acceptable? This was discussed before, in Sect. 2.5, under the heading of a-posteriori analysis. Here, the only restriction is that the number of iterations k must be taken into account, so that the value of $\hat{x}^{(k)}$ will not be misjudged. The condition of compatibility must then reduce to the form adopted by Oettli and Prager (cf. Sects. 2.4 and 2.5) when k approaches infinity. Writing

$$\begin{aligned} \hat{x}^{(k)} &= \sum_{i=0}^{k-1} (J + \Delta J)^i \, (h + \Delta h), \qquad x^{(0)} = 0 \\ &= (I - [J + \Delta J])^{-1} \, (h + \Delta h) - (I - [J + \Delta J])^{-1} \, (J + \Delta J)^k \, (h + \Delta h) \end{aligned}$$

we obtain

$$(I - [J + \Delta J]) \, \hat{x}^{(k)} = h + \Delta h - (J + \Delta J)^k \, (h + \Delta h)$$

but

$$r^{(k)} = \hat{x}^{(k)} - J\hat{x}^{(k)} - h$$

And therefore

$$r^{(k)} = \Delta J\hat{x}^{(k)} + \Delta h - (J + \Delta J)^k (h + \Delta h)$$

The last term on the right-hand side of the last equation can be simplified, when k assumes large values, to yield

$$(J + \Delta J)^k (h + \Delta h) \cong \lambda_1^k (u^1\rangle \langle v^1)\, h = \lambda_1^k \langle v^1, h\rangle\, u^1$$

where u^1 and v^1 are the eigenvector and reciprocal eigenvector of J corresponding to λ_1. The condition of compatibility of a solution $\hat{x}^{(k)}$, obtained after k iterations for the system $x^{(i+1)} = Jx^{(i)} + h$, with the uncertainty ΔJ and Δh in the data finally becomes

$$|r^{(k)}| \leqq |\Delta J|\, |\hat{x}^{(k)}| + |\Delta h| + \varrho^k\, |\langle v^1, h\rangle|\, |u^1|$$

or alternatively, due to rounding

$$|r^{(k)}| \leqq \varepsilon(|J|\, |\hat{x}^{(k)}| + |h|) + \varrho^k |\langle v^1, h\rangle|\, |u^1|$$

where ε stands for the machine precision. For values of k approaching infinity, Oettli and Prager's criterion follows directly, as a special case.

The above compatibility criterion can on the other hand be written in terms of the stoppage criterion vector $\underset{\sim}{\alpha} = \hat{x}^{(k+1)} - \hat{x}^{(k)}$. From

$$\begin{aligned}\hat{x}^{(k+1)} - \hat{x}^{(k)} &= (J + \Delta J)\, \hat{x}^{(k)} + h + \Delta h - \hat{x}^{(k)} \\ &= \Delta J\hat{x}^{(k)} + \Delta h - r^{(k)}\end{aligned}$$

we directly have that

$$|r^{(k)}| \leqq \varepsilon(|J|\, |\hat{x}^{(k)}| + |h|) + |\underset{\sim}{\alpha}|$$

which is a more practical criterion to observe and does not restrict $x^{(0)}$ to be zero as for our starting vector. Also, substituting for $r^{(k)}$ in the relation

$$\|\hat{x}^{(k)} - x\| \leqq \|\, |(I - J)^{-1}|\, |r^{(k)}|\, \|$$

results in the following bound, which is a refinement on Skeel's one, discussed in Sect. 1.5

$$\|\hat{x}^{(k)} - x\| \leqq \varepsilon \|\, |(I - J)^{-1}|\, (|J|\, |\hat{x}^{(k)}| + |h|)\| + \|\, |(I - J)^{-1}|\, |\underset{\sim}{\alpha}|\, \|$$

To apply the foregoing a-posteriori measure of $\{r^{(k)}, \hat{x}^{(k)}\}$ on an example, let us write

$$A = \begin{bmatrix} \alpha & \beta & \beta & \beta & \beta \\ \beta & \alpha & \beta & \beta & \beta \\ \beta & \beta & \alpha & \beta & \beta \\ \beta & \beta & \beta & \alpha & \beta \\ \beta & \beta & \beta & \beta & \alpha \end{bmatrix}, \qquad b = \begin{bmatrix} 1 \\ 0 \\ 0 \\ 0 \\ 0 \end{bmatrix}$$

where $\alpha = 8000.00002$ and $\beta = -1999.99998$. Here, $\|J\|$ is very near to unity with cond (A) large. The solution, by direct method of elimination, yields

$$\hat{x} = 2014.6\,(1, 1, 1, 1, 1)^T$$

and

$$r = A\hat{x} - b = 0.00146\,(1, 1, 1, 1, 1)^T$$

The residual is very high. In practice, these systems are preconditioned before attempting to solve them using the direct methods. To solve the above system using Jacobi's method, we have

$$J = \begin{bmatrix} 0 & a & a & a & a \\ a & 0 & a & a & a \\ a & a & 0 & a & a \\ a & a & a & 0 & a \\ a & a & a & a & 0 \end{bmatrix}, \qquad h = \begin{bmatrix} h_1 \\ 0 \\ 0 \\ 0 \\ 0 \end{bmatrix}$$

where $a = 2.499999969 \times 10^{-1}$; $h_1 = 1.249999997 \times 10^{-4}$. The eigenvalues of J are $-a, -a, -a, -a, 4a$. Hence $\varrho = 4a = 0.9999999875$, showing a high condition number as well as a very small rate of convergence. In fact, after 40 iterations, the results obtained were

$$\hat{x} = \begin{bmatrix} 1.079\,999\,754 \times 10^{-3} \\ 9.799\,997\,538 \times 10^{-4} \\ 9.799\,997\,538 \times 10^{-4} \\ 9.799\,997\,538 \times 10^{-4} \\ 9.799\,997\,538 \times 10^{-4} \end{bmatrix}$$

And since the correct answer, down to 10 decimal digits, comes as

$$x = \begin{bmatrix} 2\,000.000\,080 \\ 1\,999.999\,980 \\ 1\,999.999\,980 \\ 1\,999.999\,980 \\ 1\,999.999\,980 \end{bmatrix}$$

we can easily realize how far we are from the exact solution. It takes some 10^9 iterations to come any near to it. However, let us demonstrate the use of the compatibility condition at 40 iterations anyway. The residual $r^{(40)}$ is calculated to be

$$r^{(40)} = -2.499998743 \times 10^{-5}\,(1, 1, 1, 1, 1)^T$$

Also,

$$\underset{\sim}{\alpha} = \begin{bmatrix} 2.499\,998\,700 \times 10^{-5} \\ 2.499\,998\,720 \times 10^{-5} \\ 2.499\,998\,720 \times 10^{-5} \\ 2.499\,998\,720 \times 10^{-5} \\ 2.499\,998\,720 \times 10^{-5} \end{bmatrix}$$

and

$$|J|\,|\hat{x}^{(40)}| + |h| = \begin{bmatrix} 1.104\,999\,741 \times 10^{-3} \\ 1.004\,999\,741 \times 10^{-3} \\ 1.004\,999\,741 \times 10^{-3} \\ 1.004\,999\,741 \times 10^{-3} \\ 1.004\,999\,741 \times 10^{-3} \end{bmatrix}$$

For $\hat{x}^{(40)}$ to be a compatible solution at the 40th iteration, the following inequality must hold

$$2.499\,998\,743 \times 10^{-5} \begin{bmatrix} 1 \\ 1 \\ 1 \\ 1 \\ 1 \end{bmatrix} \leq 5 \times 10^{-10} \begin{bmatrix} 1.104\,999\,741 \times 10^{-3} \\ 1.004\,999\,741 \times 10^{-3} \\ 1.004\,999\,741 \times 10^{-3} \\ 1.004\,999\,741 \times 10^{-3} \\ 1.004\,999\,741 \times 10^{-3} \end{bmatrix}$$

$$+ \begin{bmatrix} 2.499\,998\,700 \times 10^{-5} \\ 2.499\,998\,720 \times 10^{-5} \\ 2.499\,998\,720 \times 10^{-5} \\ 2.499\,998\,720 \times 10^{-5} \\ 2.499\,998\,720 \times 10^{-5} \end{bmatrix} = \begin{bmatrix} 2.499\,998\,755 \times 10^{-5} \\ 2.499\,998\,770 \times 10^{-5} \\ 2.499\,998\,770 \times 10^{-5} \\ 2.499\,998\,770 \times 10^{-5} \\ 2.499\,998\,770 \times 10^{-5} \end{bmatrix}$$

This is verified in our case. Here the effect of the errors introduced by rounding is small with respect to that of error introduced by the termination of the iteration process. For large enough values of k, the opposite situation prevails, and we end up with Oettli and Prager's bound. This is clearly illustrated in Fig. 2.

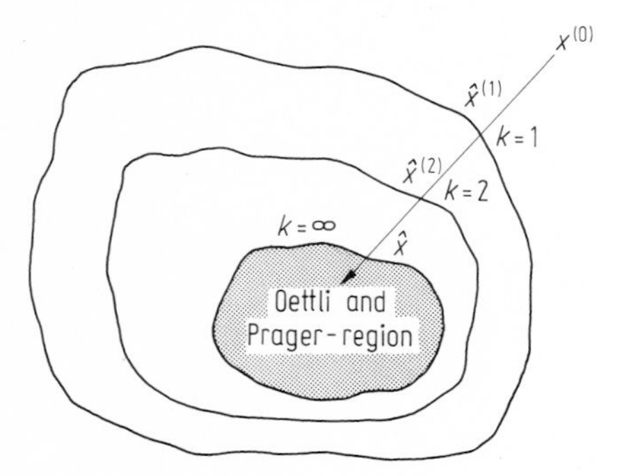

Fig. 3.2

By starting from an initial guess $x^{(0)}$, we iterate successively until we reach, at best, $\hat{x}$ when k tends to infinity. Any $\hat{x}$ in the shaded region is an admissible solution. Nevertheless, any solution $\hat{x}^{(k)}$ not lying in the shaded area is still compatible with the uncertainty ΔJ and Δh if it satisfies the foregoing criterion for $r^{(k)}$. Note that $\| |(I - J)^{-1}| \, |r^{(k)}| \|$ defines the right bound for $\|\hat{x}^{(k)} - x\|$, and yet $\hat{x}^{(k)}$ need not be an admissible solution. Indeed it would be one, if the true error in $\hat{x}^{(k)}$ is of the same order as Skeel's bound. Alternatively, $\hat{x}^{(k)}$ is an acceptable solution if the *backward error*

$$\max_i \frac{|r_i^{(k)}| - |\underset{\sim}{\alpha}_i|}{(|J| \, |\hat{x}^{(k)}| + |h|)_i}$$

is smaller than the round-off error ε.

3.5 A Method for Order Reduction

Varah (1973) suggested an interesting method for handling ill-conditioned problems, by reducing the number of equations, only choosing those equations which are best conditioned. Although his technique is different from the one described hereunder, it has indeed been at the origin of the treatment in this section. First, we will explain Varah's technique. Suppose we want to solve a system $Ax = b$ that has a high cond (A). The error in the solution due to rounding is bounded by

$$\frac{\|\Delta x\|_2}{\|x\|_2} \leqq k(n, \varepsilon) \frac{\sigma_1}{\sigma_n}$$

where σ_1 and σ_n are the largest and smallest singular values of A, their quotient being cond (A). $k(n, \varepsilon)$ is a bound for $\|\Delta A\|/\|A\|$ (cf. Sect. 1.2).

By order reduction, we mean the fact of truncating the number of equations n to a number $k < n$, having a smaller condition number. This would reduce the effect of rounding, and consequently, will improve the accuracy. However, by truncating the equations, we will have introduced a new type of error, that due to the truncation itself, since we will have dispensed with some of the information which could have served better to determine the solution. On the other hand, if both types of error present are combined and found to be smaller in magnitude than the original error, before order reduction — that is due to rounding alone — then we will have succeeded in bringing about an improvement in accuracy. Varah's technique, unfortunately, is only suitable for a special kind of problems, problems where the vector b has particular properties. To see this, the author considered singular value decomposition of A as $A = UDV^*$, where U is an orthogonal matrix of the eigenvectors of AA^*; V that of A^*A. Note that if A is normal, we have the equivalent simple diagonalization formula $A = PDP^T$. D in either case is a diagonal matrix of the singular values of A, or of the eigenvalues of A if A is normal. Now

$$UDV^*x = b$$

whence

$$x = VD^{-1}U^*b$$

To see how the equations can be truncated, the author took for his new variables the following

$$x = Vy \quad \text{and} \quad U^*b = \beta$$

thus yielding

$$Dy = \beta$$

with $y_i = \beta_i/\sigma_i$. Now if b is such that β_i/σ_i decreases as i increases (this being a restriction imposed by the author), then one can neglect the higher terms, i.e. those for which i is greater than some k. Truncating the equations to k terms, one would get

$$x^{(k)} = Vy^{(k)}$$

where

$$y^{(k)} = \sum_{i=1}^{k} \frac{\beta_i}{\sigma_i} e_i\,, \qquad e_i \text{ is the orthogonal basis.}$$

The overall error is therefore bounded by

$$\|\hat{x}^{(k)} - x\|_2 \leqq k_1(k, \varepsilon)\frac{\sigma_1}{\sigma_k} + \left[\sum_{i=k+1}^{n}\left(\frac{\beta_i}{\sigma_i}\right)^2\right]^{1/2}$$

The first term on the right-hand side accounts for the effect of round-off, whilst the second is that error due to truncation. And since the first term goes increasing with k, whereas the second decreases as k increases, there will exist some optimum value of k to choose for the error bound to be minimal. The only limitation on the technique is the fact that the quotient (β_i/σ_i) has to be decreasing as i increases. This is guaranteed for some boundary value problems which the author treated in his same paper. After all, the idea of replacing the smaller singular values by zero has already been used in solving least-squares problems; refer to Golub (1965) and Björck (1968). This procedure has been used for finding the effective rank of a matrix as well, by Peters and Wilkinson (1970).

The method described here does not impose any conditions on b, nor does it involve any approximation. Yet, it alleviates the effect of a high condition number by forming a new matrix having a lower condition number. Take for instance the example of the 5×5 matrix depicted in Sect. 3.4, here $\varrho = 0.9999999875$, forming our main obstacle as it yields a very small rate of convergence. By cancelling it from the matrix we obtain a matrix J_1 having only $-a$, $-a$, $-a$, $-a$ and 0 for eigenvalues. This speeds up the convergence. To see how this is achieved, substitute J in $x = Jx + h$ by its spectral form

$$J = \sum_{i=1}^{n} \lambda_i u^i u^{iT}$$

where u^i is the eigenvector of J corresponding to λ_i (J is taken as normal for the sake of simplicity). Hence, assuming that there is an eigenvalue of J very much near to unity in magnitude, like λ_1; i.e. $1 \geqq |\lambda_1| \geqq |\lambda_2| \geqq \dots$, then

$$x = \left(\sum_{i=2}^{n} \lambda_i u^i u^{iT} + \lambda_1 u^1 u^{1T} \right) x + h$$

or

$$(I - \lambda_1 u^1 u^{1T})\, x = \left(\sum_{i=2}^{n} \lambda_i u^i u^{iT} \right) x + h = J_1 x + h$$

Hence

$$x = (I - \lambda_1 u^1 u^{1T})^{-1} (J_1 x + h) = \left(I + \frac{\lambda_1}{1 - \lambda_1} u^1 u^{1T} \right) (J_1 x + h)$$

$$= J_1 x + \frac{\lambda_1}{1 - \lambda_1} u^1 u^{1T} J_1 x + h + \frac{\lambda_1}{1 - \lambda_1} u^1 u^{1T} h$$

The second term on the right-hand side vanishes; finally, we obtain the following recurrence relation

$$x^{(i+1)} = J_1 x^{(i)} + \left(I + \frac{\lambda_1}{1 - \lambda_1} u^1 u^{1T} \right) h$$

Here, J_1 has been rid of λ_1. With minor alterations, the above procedure can still be used with any J having values of λ_1 that possess nonlinear divisors. It can also be applied to the case where J has two greater eigenvalues λ_1 and λ_2, etc . . . Aside from this, the technique is simple, as it only requires the knowledge of λ_1 and u^1. These can be calculated using Von Mises' power method (see Carnahan, Luther and Wilkes (1969)). To apply the method on the example in Sect. 3.4, with $\lambda_1 = \varrho = 0.999999987 5$, $u^1 = (1/\sqrt{5}, 1/\sqrt{5}, 1/\sqrt{5}, 1/\sqrt{5}, 1/\sqrt{5})^T$ the new system $x = J_1 x + h^1$ has

$$J_1 = \begin{bmatrix} \alpha & \beta & \beta & \beta & \beta \\ \beta & \alpha & \beta & \beta & \beta \\ \beta & \beta & \alpha & \beta & \beta \\ \beta & \beta & \beta & \alpha & \beta \\ \beta & \beta & \beta & \beta & \alpha \end{bmatrix}, \qquad h^1 = \begin{bmatrix} 2\,000.000\,100 \\ 1\,999.999\,975 \\ 1\,999.999\,975 \\ 1\,999.999\,975 \\ 1\,999.999\,975 \end{bmatrix}$$

$\alpha = -0.199999998$, $\beta = 0.049999999$. J_1 has eigenvalues of order 4 ($\lambda = -0.249999997$) and a zero eigenvalue. It will thus converge very quickly during iteration. Here, one should note that h^1 approximates very well the solution, and therefore x is reached after only very few iterations. Furthermore, the condition number will have been reduced, for cond $(I - J + \lambda_1 u^1 u^{1T}) \leq$ cond $(I - J)$. Further, especially for matrix A having property A, using the spectral norm, do we have

$$\text{Cond}\,(I - J_1) = \frac{1 + \lambda_2}{1 - \lambda_2}$$
$$\leq \frac{1 + \lambda_1}{1 - \lambda_1} = \text{cond}\,(I - J)$$

In brief, the above technique relies on the removal of the contamination caused by large eigenvalues, both in the rate of convergence and in the condition number. Further, it relies on the power method in deflating the matrix J. This does not mean that we have replaced one iterative method by another when first finding the eigenvalues, for the original system's accuracy depends on λ_1, whereas the accuracy of the power method when used to find λ_1 depends on the quotient $|\lambda_i/\lambda_1|$, i.e. on the better separation of the eigenvalues.

If a series of *deflation procedures* is carried out successively to remove all of the larger eigenvalues (or, alternatively, all absurd equations), we will ultimately wind up with the system

$$y_m^{(i+1)} = \tilde{J}_{m,m} y_m^{(i)} + \tilde{h}_m$$

where $\tilde{J}$ is of a lesser dimension than J. This will be achieved through the use of a *proper* congruent transformation $x = Py$. Furthermore, the last terms $y_{m+1}, \dots, y_n$ can be obtained by direct inspection, where $n - m$ stands for the number of the removed equations.

3.6 Methods of Iterative Refinement

Until now, we have used the residual r corresponding to an approximate solution $\hat{x}$ of the equations $Ax = b$, namely

$$r = A\hat{x} - b$$

for two purposes

1. As an indicator of the accuracy of the solution $\hat{x}$. In general, the smaller $\|r\|$, the better is $\hat{x}$ or the nearer it is to x.
2. As an indicator of the compatibility of $\hat{x}$ with the uncertainties in A and b. $\hat{x}$ is termed an admissible solution according to whether r satisfies or not some a-posteriori criterion.

In this section we will demonstrate a new application, namely:

3. Solving $Ax = b$ iteratively by successively calculating the residual $r^{(i)}$ after every iteration, starting from an approximate solution $x^{(0)}$. This class of methods is

termed that of *iterative refinement*. Herewith, $r^{(i)}$ is used either to solve $Ax = b$, or to improve on an approximate solution $x^{(0)}$. These methods can also be used to find the inverse A^{-1} of a matrix starting from a first approximation $B^{(0)}$.

For example, the steps used to achieve an improvement on the solution $\hat{x}$ of the equation $Ax = b$ would be:

a) Compute $r^{(i)} = Ax^{(i)} - b$
b) Solve the system $Ay^{(i)} = r^{(i)}$
c) $x^{(i+1)} = x^{(i)} - y^{(i)}$

Then $x^{(i)}$ will approach the solution $x = A^{-1}b$ as i approaches infinity. Before we proceed with the proof, it is worthy of notice that $r^{(i)}$ is calculated using double-precision floating-point arithmetic, see also exercise 3.19. Solving $Ay^{(i)} = r^{(i)}$ using single precision arithmetic is synonymous to finding $y^{(i)} = (A + \Delta A)^{-1}\, r^{(i)}$. The proof proper comes as

$$\begin{aligned} x^{(k+1)} &= x^{(k)} - (A + \Delta A)^{-1}\,(Ax^{(k)} - b) \\ &= (I - (A + \Delta A)^{-1}\, A)\,(x^{(k)} - x) + x \end{aligned}$$

This is equivalent to saying that

$$x^{(k+1)} - x = (I - (A + \Delta A)^{-1}\, A)\,(x^{(k)} - x)$$

whence

$$x^{(k+1)} - x = (I - (A + \Delta A)^{-1}\, A)^{k+1}\,(x^{(0)} - x)$$

but since usually (see exercise 3.20)

$$\|I - (A + \Delta A)^{-1}\, A\| \ll 1\,,$$

$x^{(k)}$ will approach the solution x as k approaches infinity.

Note that, since step b is usually repeated more than just once, one can store an estimate B of A^{-1} and carry on with the operations using only matrix multiplication. In fact, some authors prefer to use the recurrence relations in the form

$$\begin{aligned} x^{(k+1)} &= x^{(k)} - B(Ax^{(k)} - b) \\ &= [I - BA]\, x^{(k)} + Bb \end{aligned}$$

similar to the iterative equations in Sect. 3.1. Therefore, the rate of convergence, noted R, becomes

$$R = -\frac{1}{k} \ln \|(I - BA)^k\|$$

(cf. Sect. 3.3). Likewise, the asymptotic rate of convergence is given by

$$-\ln\,[\varrho(I - BA)]$$

which is a relatively high rate of convergence in comparison with previous ones obtained with other iterative methods, since

$$\|I - BA\| \ll 1$$

This does not imply that this method is faster in the absolute, since it necessitates an a-priori estimate of the matrix B.

Similarly to the analysis in Sect. 3.4, we will ask, if

$$\|x^{(k+1)} - x^{(k)}\| < \alpha$$

then how remote from x is $x^{(k)}$? One can indeed easily deduce that

$$\|x^{(k)} - x\| \leqq \frac{\alpha}{1 - \varrho}$$

Also, if $x^{(0)}$ is an approximate solution to $Ax = b$, then how far is x from $x^{(0)}$? This depends, obviously, on the value of the initial residual $Ax^{(0)} - b$. From the relation

$$x^{(k)} - x^{(0)} = \left(-\sum_{i=0}^{k-1} (I - BA)^i\right) B(Ax^{(0)} - b)$$

it follows that, as $k \to \infty$, the final solution x satisfies

$$|x - x^{(0)}| \leqq (I - |I - BA|)^{-1} |B| \, |Ax^{(0)} - b|$$

componentwise. This simply means that the solution $x^{(0)}$ approximates x by a factor proportional to B. Therefore, the more ill-conditioned is A the farther is $x^{(0)}$ from x. This is indeed to be expected. The above bound in the present form is due to Yamamoto (1981).

To exemplify the method of iterative refinement, let us consider the example in Hamming (1971), discussed in Sect. 1.5, namely: solve

$$A = \begin{bmatrix} 3 & 2 & 1 \\ 2 & 2\varepsilon & 2\varepsilon \\ 1 & 2\varepsilon & -\varepsilon \end{bmatrix}, \qquad b = \begin{bmatrix} 3 + 3\varepsilon \\ 6\varepsilon \\ 2\varepsilon \end{bmatrix}$$

which has for an exact solution

$$x = (\varepsilon, 1, 1)^T$$

For a value of ε equal to 10^{-9}, using a ten-digit-mantissa, floating-point machine and direct elimination, the solution was found to be

$$\hat{x} = (1.166666667 \times 10^{-9}, 1.024351388, 9.512972235 \times 10^{-1})^T$$

which is a very inaccurate result. Instead, use of the method of iterative refinement, while still solving $Ay^{(i)} = r^{(i)}$ by direct elimination, yields the solution after seven iterations. The results of these are tabulated in sequence in Table 1.

Another example would be the improvement of an approximate inverse $B^{(0)}$ to A. First, we form the residual

$$R^{(0)} = I - AB^{(0)}; \qquad \|R^{(0)}\| \ll 1$$

Then, we construct the two sequences, according to Demidovich and Maron (1973, p. 316), as

$$\begin{aligned} R^{(1)} &= I - AB^{(1)} & B^{(1)} &= B^{(0)} + B^{(0)}R^{(0)} \\ R^{(2)} &= I - AB^{(2)} & B^{(2)} &= B^{(1)} + B^{(1)}R^{(1)} \\ &\vdots & &\vdots \end{aligned}$$

Then, as k approaches infinity, $B^{(k)}$ will approach A^{-1}. This result can be proven using the following relation

$$\begin{aligned} B^{(1)} &= B^{(0)}(I + R^{(0)}) \\ &= A^{-1}(I - R^{(0)})(I + R^{(0)}) \\ &= A^{-1}(I - [R^{(0)}]^2) \\ B^{(2)} &= B^{(1)}(I + R^{(1)}) \\ &= A^{-1}(I - [R^{(0)}]^2)(2I - I + [R^{(0)}]^2) \\ &= A^{-1}(I - [R^{(0)}]^4) \end{aligned}$$

In general

$$B^{(k)} = A^{-1}(I - [R^{(0)}]^{2^k})$$

Therefrom, it follows that $B^{(k)}$ approaches A^{-1} — as k approaches infinity — very rapidly. The reader may experiment this method on the following matrix:

$$A = \begin{bmatrix} 1/2 & 1/3 & 1/4 \\ 1/3 & 1/4 & 1/5 \\ 1/4 & 1/5 & 1/6 \end{bmatrix}$$

of which the approximate inverse $B^{(0)}$ is

$$B^{(0)} = \begin{bmatrix} 7.199\,999\,783 \times 10^1 & -2.399\,999\,920 \times 10^2 & 1.799\,999\,935 \times 10^2 \\ -2.399\,999\,919 \times 10^2 & 8.999\,999\,698 \times 10^2 & -7.199\,999\,756 \times 10^2 \\ 1.799\,999\,935 \times 10^2 & -7.199\,999\,756 \times 10^2 & 5.999\,999\,803 \times 10^2 \end{bmatrix}$$

Table 1

n	x_1	x_2	x_3	r_1	r_2	r_3
0	$1.166666667 \times 10^{-9}$	1.024351388	$9.512972235 \times 10^{-1}$	0	$2.84630557 \times 10^{-10}$	$2.640722195 \times 10^{-10}$
1	$1.003333334 \times 10^{-9}$	$9.991606444 \times 10^{-1}$	1.001678711	-1.89999×10^{-10}	$8.3453788 \times 10^{-12}$	-2.40882×10^{-14}
2	$9.998663340 \times 10^{-10}$	1.000028931	$9.999421374 \times 10^{-1}$	$-6.004010 \times 10^{-10}$	$-3.251952 \times 10^{-13}$	-1.79414×10^{-14}
3	$1.000006667 \times 10^{-9}$	$9.999990028 \times 10^{-1}$	1.000001994	$-3.999799990 \times 10^{-10}$	1.532760×10^{-14}	2.6786×10^{-15}
4	10^{-9}	1.000000034	$9.999999313 \times 10^{-1}$	-7.0×10^{-10}	-6.94×10^{-17}	1.367×10^{-16}
5	$1.000000003 \times 10^{-9}$	$9.999999988 \times 10^{-1}$	1.000000002	$-3.999999910 \times 10^{-10}$	7.6×10^{-18}	-1.4×10^{-18}
6	10^{-9}	1.000000000	$9.999999999 \times 10^{-1}$	-10^{-10}	-2×10^{-19}	10^{-19}
7	10^{-9}	1	1	0	0	0

Usually, the method of iterative refinement is not applied indefinitely. One iteration is sufficient to bring about noticeable improvement. For example, for

$$A = \begin{bmatrix} 33 & 16 & 72 \\ -24 & -10 & -57 \\ -8 & -4 & -17 \end{bmatrix}$$

which has for an exact inverse the following

$$A^{-1} = \begin{bmatrix} -29/3 & -8/3 & -32 \\ 8 & 5/2 & 51/2 \\ 8/3 & 2/3 & 9 \end{bmatrix}$$

the approximate inverse obtained on a 10-digit machine was

$$B^{(0)} = \begin{bmatrix} -9.666\,666\,881 & -2.666\,666\,726 & -32.000\,000\,71 \\ 8.000\,000\,167 & 2.500\,000\,046 & 25.500\,000\,55 \\ 2.666\,666\,728 & 0.666\,666\,683\,6 & 9.000\,000\,203 \end{bmatrix}$$

which is correct to seven digits. One iteration of refinement yields

$$B^{(1)} = \begin{bmatrix} -9.666\,666\,667 & -2.666\,666\,667 & -32.000\,000\,000\,0 \\ 8.000\,000\,000 & 2.500\,000\,000 & 25.500\,000\,000\,0 \\ 2.666\,666\,667 & 0.666\,666\,666\,7 & 9.000\,000\,000\,0 \end{bmatrix}$$

which is correct up to ten digits.

Usually, the residual is computed using double-precision — even if it is subsequently stored in single-precision form to guarantee the accumulation of inner products of vectors. Accumulation is vital, especially if cond (A) is large in value. Wilkinson (1965, p. 260) gave accuracy estimates for the computed solutions after every iteration, together with the computed residuals. He also provided a rule of thumb to guarantee convergence. According to him, the process will in general succeed if

$$n^k \varepsilon \, \|A^{-1}\| < 1$$

where n is the dimension of A and ε the machine precision.

Another very popular method for iterative refinement, sometimes misnamed in the literature as the method of relaxation (cf. Salvadori and Baron, 1966), makes use of the residual to update x without solving the whole system at each iteration. It is mainly suitable for diagonal dominant matrices, but has still the advantage of being easily implementable on a calculator having only the $+$, $-$, $\times$, $\div$ operations. In return, convergence with this method is slow. It relies on the reduction of the numerically largest residual to zero at each step; and terminates when all residuals of the last equation vanish.

Let the equations be written in the form

$$Ax^{(0)} - b = r^{(0)}$$

choosing $x^{(1)}$ as

$$x^{(1)} = x^{(0)} + \Delta x^{(0)}$$

such that when $Ax^{(1)} - b = r^{(1)}$, then

$$\|r^{(1)}\| \leqq \|r^{(0)}\|$$

The problem then becomes one of determining a suitable value of $\Delta x^{(0)}$. From the above relations, we have that

$$\begin{aligned} r^{(1)} = Ax^{(1)} - b &= A(x^{(0)} + \Delta x^{(0)}) - b \\ &= A\,\Delta x^{(0)} + r^{(0)} \end{aligned}$$

Therefore, a good choice of $\Delta x^{(0)}$ is one which ensures that

$$\|A\,\Delta x^{(0)} + r^{(0)}\| \leqq \|r^{(0)}\|$$

Let $r_k^{(0)}$ be the largest element in magnitude in $r^{(0)}$. Choose $\Delta x^{(0)}$ as follows

$$\Delta x^{(0)} = (0, 0, \ldots, 0, \Delta x_k^{(0)}, 0, \ldots, 0)^T$$

so that

$$\Delta x_k^{(0)} = -\frac{r_k^{(0)}}{a_{kk}}$$

Hence, based on the definition of the l_1-norm, we have that

$$\begin{aligned} \|A\,\Delta x^{(0)} + r^{(0)}\| &= \sum_{i \neq k} \left| r_i^{(0)} - \frac{a_{ik} r_k^{(0)}}{a_{kk}} \right| \\ &= \sum_{i \neq k} \left| r_i^{(0)} - \frac{a_{ik} r_k^{(0)}}{a_{kk}} \right| + |r_k^{(0)}| - |r_k^{(0)}| \\ &\leq \sum_{i \neq k} |r_i^{(0)}| + |r_k^{(0)}| \sum_{i \neq k} \left| \frac{a_{ik}}{a_{kk}} \right| + |r_k^{(0)}| - |r_k^{(0)}| \\ &= \|r^{(0)}\| - |r_k^{(0)}| \left(1 - \sum_{i \neq k} \left| \frac{a_{ik}}{a_{kk}} \right| \right) \end{aligned}$$

But if A is diagonal dominant, the last bracket on the right-hand side is positive and

$$\|A\,\Delta x^{(0)} + r^{(0)}\| \leqq \|r^{(0)}\|$$

Hence the sequence $\|r^{(k)}\|$ is monotonic decreasing and bounded below, thus converging to a minimum value of $\|Ax - b\|$. To illustrate the method on an example, consider

$$\begin{bmatrix} 4 & -1 & -1 & 0 \\ -1 & 4 & 0 & -1 \\ -1 & 0 & 4 & -1 \\ 0 & -1 & -1 & 4 \end{bmatrix} \begin{bmatrix} x_1 \\ x_2 \\ x_3 \\ x_4 \end{bmatrix} - \begin{bmatrix} 0 \\ 0 \\ 1000 \\ 1000 \end{bmatrix} = r$$

Choosing $x^{(0)} = 0$ yields

$$r^{(0)} = (0, 0, -1000, -1000)^T$$

Next, choose

$$x_1 = 0\,, \qquad x_2 = 0\,, \qquad x_3 = 0 + (1000/4) = 250\,, \qquad x_4 = 0$$

giving

$$r^{(1)} = (-250, 0, 0, -1250)^T$$

The next choice of x_i should then be

$$x_1 = 0\,, \qquad x_2 = 0\,, \qquad x_3 = 250\,, \qquad x_4 = 0 + (1250/4) = 312.5$$

giving

$$r^{(2)} = (-250, -312.5, -312.5, 0)^T$$

Next choose

$$x_1 = 0\,, \qquad x_2 = 0\,, \qquad x_3 = 250 + (312.5/4) = 328.1\,, \qquad x_4 = 312.5$$

giving

$$r^{(3)} = (-328.1, -312.5, 0, -78.1)^T$$

Then choose

$$x_1 = 0 + (328.1/4) = 82\,, \qquad x_2 = 0\,, \qquad x_3 = 328.1\,, \qquad x_4 = 312.5$$

giving

$$r^{(4)} = (0, -394.5, -82, -78.1)^T$$

Next, choose

$$x_1 = 82\,, \qquad x_2 = 0 + (394.5/4) = 98.6\,, \qquad x_3 = 328.1\,, \qquad x_4 = 312.5$$

giving

$$r^{(5)} = (-98.6, 0, -82, -176.7)^T$$

and so forth. It takes some 20 iterations to reach the solution

$$x = (125, 125, 375, 375)^T$$

against 13 iterations for the Gauss-Siedel method.

Although convergence is slow when using this method, it appears nevertheless, that the method is very tractable when dealing with small order systems: a desk calculator could do the job perfectly, there being no need for programming. The method also appears very useful in cases where only few of the unknowns need correction. And since A is diagonal dominant, the variable coinciding horizontally with the largest-magnitude residual is the one to start with.

3.7 Case of Interval Coefficients

Iterative methods for solving the system $Ax = b$ were previously shown to be divided into two categories:

1. The equations $Ax = b$ are written in the recurrence form

   $$x^{(k+1)} = Jx^{(k)} + h$$

   whereby starting from an initial vector $x^{(0)}$, the iteration variable $x^{(k)}$ approaches x as k approaches infinity. This is guaranteed if $\varrho(J) < 1$.
2. Starting from an approximate solution $x^{(0)}$, a better solution $x^{(1)}$ is obtained using the relation $x^{(1)} = x^{(0)} - B(Ax^{(0)} - b) = (I - BA)\, x^{(0)} + Bb$, where B is an approximate inverse of A; i.e. $\|I - BA\| < 1$. This form resembles Jacobi's, see Sect. 3.1; after many iterations, $x^{(k)}$ approaches x. Unlike category 1, the method does not necessitate that A be diagonal dominant, for instance (i.e. $\varrho(J) < 1$). Another advantage alike category one, is that iteration will still converge no matter what starting vector $x^{(0)}$ we choose (see Sect. 3.6). Still, the method has a major drawback, which is the necessity of knowing an approximate inverse B of the matrix A. For this reason, it is only used to improve on an already existing solution $x^{(0)}$ obtained by direct methods. This is why it is referred to as the iterative refinement technique.

Both methods can on the other hand be used for the solution of linear interval equations $A^I x = b^I$. In this case, B represents the inverse of the mid-point matrix of A^I, i.e. $B = (A^c)^{-1}$. We shall not indulge in a detailed discussion of the application of the first category of methods to interval equations, since the case resembles that with A fixed, except for the fact that here, the equations are rewritten in the interval recurrence form

$$x^{I(k+1)} = J^I x^{I(k)} + h^I$$

with $J^I = [\underline{J}, \overline{J}]$ and $h^I = [\underline{h}, \overline{h}]$. Mayer (1968, 1970) has shown that the above interval iteration, for any starting value $x^{I(0)}$, converges to a fixed point of the equation

$$x^I = J^I x^I + h^I$$

if and only if $\varrho(J^I) < 1$ (cf. also Alefeld and Herzberger (1983, p. 190)).

If $\varrho(J^I) \not< 1$, the second method can be used instead of the first. The reader may have already noticed that this method is similar in some respects to Hansen and Smith's second method (1967), seen in Sect. 2.3, and where the authors suggested a solution by Gauss elimination of the interval equations

$$BA^I z^I = Bb^I$$

obtaining $z^I \supseteq x^I$. In the iterative form, the above equations become

$$z^{I(k+1)} = (I - BA^I)\, z^{I(k)} + Bb^I = E^I z^{I(k)} + Bb^I$$

Here E^I is an error interval matrix (see Sect. 2.3) having $\|E^I\| < 1$, a fact which implies convergence. Note that $z^{I(0)}$ must be so chosen as to enclose x^I. Meanwhile, increasing the number of iterations will reduce the overestimation error and tighten the bound between $z^{I(k)}$ and x^I until a value of z^I is reached that satisfies

$$z^I = E^I z^I + Bb^I$$

The iterations are therefore written in the following concise form, from Moore (1979, p. 61), as

$$z^{I(k+1)} = \{E^I z^{I(k)} + Bb^I\} \cap z^{I(k)}, \qquad k = 0, 1, 2, \ldots$$

with

$$z_i^{I(0)} = [-1, 1] \frac{\|Bb^I\|}{1 - \|E^I\|}, \qquad i = 1, 2, \ldots, n$$

Here, $z^{I(k)}$ — with $k = 1, 2, \ldots$ — define a nested sequence of interval vectors containing the unique solution z^I.

To illustrate the above technique, let us solve the following example:

$$A^I = \begin{bmatrix} [1.9, 2.1] & [0.9, 1.1] \\ [0.9, 1.1] & [0.9, 1.1] \end{bmatrix}, \qquad b^I = \begin{bmatrix} [2.9, 3.1] \\ [1.9, 2.1] \end{bmatrix}$$

By solving the equations $A^I x = b^I$, we calculate in fact the minimum interval vector x^I that encloses the set $X = \{x : Ax = b,\ A \in A^I,\ b \in b^I\}$. This is achieved correctly using the method in Sect. 2.4, or by Hansen's third method, sec. 2.3.

In using the above iterative method, one first obtains

$$B = \begin{bmatrix} 1 & -1 \\ -1 & 2 \end{bmatrix}, \qquad Bb^I = \begin{bmatrix} [0.8, 1.2] \\ [0.7, 1.3] \end{bmatrix}, \qquad E^I = \begin{bmatrix} [-0.2, 0.2] & [-0.2, 0.2] \\ [-0.3, 0.3] & [-0.3, 0.3] \end{bmatrix}$$

And by choosing

$$z_{1,2}^{I(0)} = [-1, 1] \frac{1.3}{1 - 0.6} = [-3.25, 3.25]$$

we obtain, using a two-decimal-digit rounded interval arithmetic:

$$z^{I(1)} = \begin{bmatrix} [-0.5, 2.5] \\ [-1.25, 3.25] \end{bmatrix}$$

$$z^{I(2)} = \begin{bmatrix} [-0.35, 2.35] \\ [-1.03, 3.03] \end{bmatrix}$$

$$z^{I(3)} = \begin{bmatrix} [-0.28, 2.28] \\ [-0.92, 2.92] \end{bmatrix}$$

$$z^{I(4)} = \begin{bmatrix} [-0.24, 2.24] \\ [-0.86, 2.86] \end{bmatrix}$$

$$z^{I(5)} = \begin{bmatrix} [-0.22, 2.22] \\ [-0.83, 2.83] \end{bmatrix}$$

The iterations are seen to converge; they will ultimately yield

$$z^{I(9)} = z^{I(10)} = z^I = \begin{bmatrix} [-0.2, 2.2] \\ [-0.8, 2.8] \end{bmatrix} \supset x^I$$

The result satisfies exactly the equation $z^I = E^I z^I + Bb^I$.

The above technique can also be applied in the solution of non-interval equations; here they will account for rounding errors. Moore (1979, p. 61) produced the following example

$$A = \begin{bmatrix} 3 & 1 \\ 3 & 2 \end{bmatrix}, \qquad b = \begin{bmatrix} 1 \\ 0 \end{bmatrix}$$

Hence

$$B = \begin{bmatrix} 0.6 & -0.3 \\ -1 & 1 \end{bmatrix}, \quad Bb = \begin{bmatrix} 0.6 \\ -1 \end{bmatrix}, \quad E = \begin{bmatrix} 0.1 & 0 \\ 0 & 0 \end{bmatrix}, \quad x^{(0)} = \begin{bmatrix} [-1.12, 1.12] \\ [-1.12, 1.12] \end{bmatrix}$$

Then, we can get

$$x^{(1)} = \begin{bmatrix} [0.488, 0.712] \\ -1 \end{bmatrix}$$

$$x^{(2)} = \begin{bmatrix} [0.648, 0.672] \\ -1 \end{bmatrix}$$

$$x^{(3)} = \begin{bmatrix} [0.664, 0.668] \\ -1 \end{bmatrix}$$

$$x^{(4)} = \begin{bmatrix} [0.666, 0.667] \\ -1 \end{bmatrix}$$

$x^{(4)}$ is the exact solution, within the range of rounding errors. Usually the equations $Ax = b$ are first solved using Gauss elimination, or a variant of it like the Gauss-Jordan method; the approximate solution $x^{(0)}$ is expressed in rounded interval arithmetic of a large width. The interval refinement method is then applied. The reader may refer to Rump and Kaucher (1980) for a treatment of the topic.

Still, the above method for obtaining z^I can be improved on to yield a tighter bound y^I. This has been achieved by Gay (1982), who succeeded in finding an interval vector y^I enclosing x^I while being of smaller width than z^I, i.e.

$$z^I \supseteq y^I \supseteq x^I$$

In his analysis, Gay made use of a theorem introduced by Miranda (1941) stating that if $g(y)$ is a vector-valued function having the property that $g_i(y) \leqq 0$ for some $y_i = \underline{y}_i$, and $g_i(y) \geqq 0$ for some $y_i = \bar{y}_i$, then there exists a y at which $g(y) = 0$. To apply this result in finding a minimum y^I enclosing the set $X = \{x : Ax = b, A \in A^I, b \in b^I\}$, we should note that any y^I enclosing X satisfies the relation $A^I y^I \supseteq b^I$. Likewise, the two inequalities

$$\sup \left(C_{ii}\underline{y}_i + \sum_{j \neq i} C_{ij} y_j \right) \leq \underline{d}_i$$

$$\inf \left(C_{ii}\bar{y}_i + \sum_{j \neq i} C_{ij} y_j \right) \geq \bar{d}_i$$

where $C = BA^I$ and $d = Bb^I$, define the set Y given by

$$Y = \{y : BAy = Bb, A \in A^I, b \in b^I\}$$

The above result immediately follows from $BA^I y^I \supseteq Bb^I$. And by defining $g(y) = B(Ay - b)$, we have $g_i(y) \leqq 0$ when $y_i = \underline{y}_i$ and $g_i(y) \geqq 0$ for $y_i = \bar{y}_i$. From this, it follows that a certain y exists at which $g(y) = B(Ay - b) = 0$. But since B is nonsingular, then $Ay = b$, i.e. Y, whose bounds are defined above, is a solution to the equations, defining the smallest set Y enclosing X. This further suggests that we proceed with the iterations using the interval-vector valued function given by

$$M_i(y^I) = \left(d_i^I - \sum_{j \neq i} C_{ij}^I y_j^I \right) \Big/ C_{ii}^I$$

while iterating on the Jacobi-like form

$$y^{I(k+1)} = M(y^{I(k)})$$

or on the Gauss-Siedel form

$$y_j^{I(k+1)} = \Big(d_j^I - \sum_{k<j} C_{jk}^I y_k^{I(k+1)} - \sum_{k>j} C_{jk}^I y_k^{I(k)}\Big) \Big/ C_{ii}^I$$

which has a higher convergence rate.

Gay further noticed that z^I, y^I and x^I all differ from one another in the sense of the metric q (see Sect. 2.3) by no more than $0(r) \cdot w(x^I)$; r being set equal to $\sup \|E^I\|$. But noticing that $\|E^I\| \approx w(E^I)$, we realize that, in fact, Miller (1972) thought of a somewhat similar bound when deducing that z^I, obtained from $BA^I z = Bb^I$ by Gauss-Jordan, differs from x^I by no more than $0[w(A^I)]^2$. Moore (1966) mentioned a similar result while analysing Hansen's first method (see Sect. 2.3).

Lastly, as our method is iterative, one must introduce an accuracy test relating two successive iterations, a test upon the success of which the process is terminated. The author suggested the tolerance criterion

$$\frac{q(y^{I(k)}, y^{I(k-1)})}{w(y^{I(k)})} \leq \alpha \text{ (prescribed)}$$

But now the question arises as to how far $y^{I(k)}$ is from x^I. One can easily see, through a similar analysis to the one in Sect. 3.4, that since

$$y^I - y^{I(k)} = (I - E^I)^{-1} E^I(y^{I(k)} - y^{I(k-1)})$$

and

$$y^I \supseteq x^I$$

then

$$\frac{q(y^{I(k)}, x^I)}{w(y^{I(k)})} \leq \frac{\alpha r}{1 - r}$$

To illustrate the author's method on an example, we will borrow one from his paper (Gay (1982)), namely

$$A^I = \begin{bmatrix} [2, 4] & [-1, 1] \\ [-1, 1] & [2, 4] \end{bmatrix}, \qquad b^I = \begin{bmatrix} [0, 2] \\ [0, 2] \end{bmatrix}$$

The set X is represented by the shaded area in Fig. 3 below.

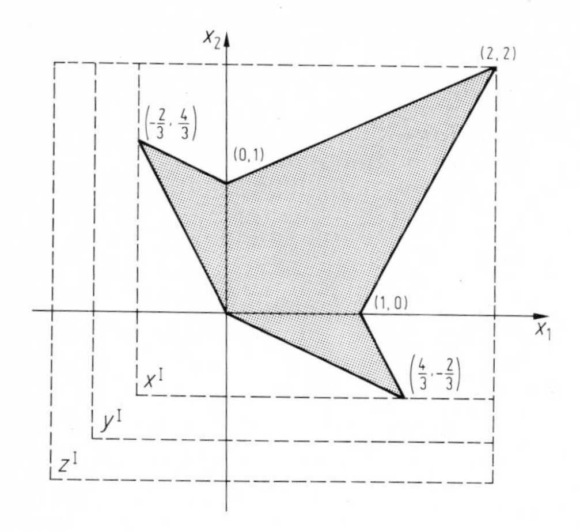

Fig. 3.3

First, the region $X = \{x: Ax = b,\ A \in A^I,\ b \in b^I\}$ is determined using the method in Sect. 2.4, wherefrom it follows that

$$x^I = \begin{bmatrix} [-2/3, 2] \\ [-2/3, 2] \end{bmatrix}$$

Then z^I is calculated using Hansen's iterative method as explained in the present section. One first gets

$$B = \begin{bmatrix} 1/3 & 0 \\ 0 & 1/3 \end{bmatrix}, \qquad Bb^I = \begin{bmatrix} [0, 2/3] \\ [0, 2/3] \end{bmatrix}, \qquad E^I = \begin{bmatrix} [-1/3, 1/3] & [-1/3, 1/3] \\ [-1/3, 1/3] & [-1/3, 1/3] \end{bmatrix}$$

and

$$z_{1,2}^{I(0)} = [-1, 1]\frac{\|Bb^I\|}{1 - \|E^I\|} = [-2, 2]$$

Then

$$z^{I(1)} = \begin{bmatrix} [-4/3, 2] \\ [-4/3, 2] \end{bmatrix}$$

$$z^{I(2)} = \begin{bmatrix} [-4/3, 2] \\ [-4/3, 2] \end{bmatrix}$$

convergence being rapidly observed, and

$$z^I = \begin{bmatrix} [-4/3, 2] \\ [-4/3, 2] \end{bmatrix}$$

Now for y^I, we use Gay's recurrence formula

$$M_{1,2}(y^{I(1)}) = ([0, 2/3] - [-1/3, 1/3]\,[-2, 2])/[2/3, 4/3]$$
$$= [-1, 2]$$

$M_{1,2}(y^{I(2)})$ will be given in a similar fashion, and therefore

$$y^I = \begin{bmatrix} [-1, 2] \\ [-1, 2] \end{bmatrix}$$

from which we have that

$$z^I \supset y^I \supset x^I$$

Exercises 3

1. Show that all block tridiagonal matrices

$$\begin{bmatrix} D_1 & A_{12} & & & \\ A_{21} & D_2 & A_{23} & & \\ & \ddots & \ddots & \ddots & \\ & & \ddots & \ddots & A_{N-1,N} \\ & & & A_{N,N-1} & D_N \end{bmatrix}$$

with D_i diagonal, $A_{ij} \neq 0$, have property A.

2. Solve by iteration the following

$$\begin{bmatrix} -4 & -1 & -1 & 0 \\ -1 & 4 & 0 & -1 \\ -1 & 0 & 4 & -1 \\ 0 & -1 & -1 & 4 \end{bmatrix} \begin{bmatrix} x_1 \\ x_2 \\ x_3 \\ x_4 \end{bmatrix} = \begin{bmatrix} 0 \\ 0 \\ 1\,000 \\ 1\,000 \end{bmatrix}$$

Choose $x^{(0)} = \underset{\sim}{0}$.

3. Prove Gershgorin's theorem: The eigenvalues of A lie in the union of the discs given by

$$|\lambda - a_{ii}| \leq \sum_{j=1,\, j\neq i}^{n} |a_{ij}|$$

Hint: $Au = \lambda u$ is $(\lambda - a_{ii})\, u_i = \sum_{j=1,\, j\neq i} a_{ij}u_j;\ i = 1, \ldots n$

Show how to use this theorem to prove that a diagonal dominant matrix is nonsingular.

4. Show that $\varrho[(D + L)^{-1}\, U] < 1$, where L and U are the lower and upper parts of A; D contains its diagonal entries. Here

$$\sum_{j\neq i} \left|\frac{a_{ij}}{a_{ii}}\right| < 1 \qquad (A \text{ is diagonal dominant})$$

Hint: consider the equation $\left|(D + L) - \frac{U}{\lambda}\right| = 0$ in λ. Show that values of λ such that $|\lambda| \geqq 1$ do not satisfy it, since $(D + L) - U/\lambda$ is diagonal dominant.

5. Show that

$$\|(D + L)^{-1} U\| \leqq \|D^{-1}(L + U)\|$$

by considering

$$|(I - D^{-1}L)^{-1}| \leqq (I - |D^{-1}L|)^{-1}$$

where $|L|$ is the matrix of absolute values of the elements of L. This should show that Gauss-Siedel is at least as fast as the Jacobi method.

6. If: $x^{(i+1)} = Hx^{(i)} + h$, show that $x^{(i+1)} - x = H^{i+1}(x^{(0)} - x)$. And if $x^{(i+1)} = Hx^{(i)} + h + \varepsilon$ (ε being a computational error), show that $x^{(i+1)} - x = H^{i+1}(x^{(0)} - x) + \sum_{i=0}^{k} H^{k-i}\varepsilon_i$

Hence, obtain a bound for $\|x^{(i+1)} - x\|$.

7. Show that if $\|J\| < 1$, then we have, under different norms:

1. $\sum_{j=1, j\neq i}^{n} \frac{|a_{ij}|}{|a_{ii}|} < 1$

2. $\sum_{i=1, i\neq j} \frac{|a_{ij}|}{|a_{ii}|} < 1$

3. $\sum_{\substack{i,j=1 \\ j\neq i}}^{n} \frac{|a_{ij}|^2}{|a_{ii}|^2} < 1$

Specify the norm used. Also, by substituting $x_i = p_i z_i$ in the expression $Ax = b$; p_i being some positive numbers, show that the system in z_i is convergent if the original system in x_i is. By taking $p_i = 1/|a_{ii}|$, show that any of the following conditions can be sufficient for convergence of the Jacobi:

a. $\sum_{\substack{j=1 \\ j\neq i}}^{n} \frac{|a_{ij}|}{|a_{jj}|} < 1$

b. $\sum_{\substack{i=1 \\ i\neq j}}^{n} \frac{|a_{ij}|}{|a_{jj}|} < 1$

c. $\sum_{\substack{i,j=1 \\ i\neq j}}^{n} \frac{|a_{ij}|^2}{|a_{jj}|^2} < 1$

Show that for A diagonal dominant, conditions 1 and b hold.

8. Show that the Gauss-Siedel method is convergent for positive definite matrices A. Hint: $G = -(D + L)^{-1} U = (Q - I)(Q + I)^{-1}$. Here $Q = A^{-1}(2[D + L] - A)$, whence $\lambda(G) = (\mu(Q) - 1)/(\mu(Q) + 1)$. Obtain $|\lambda(G)|$ by letting $\mu(Q) = \mathrm{Re}\,\mu + i\,\mathrm{Im}\,\mu$, knowing that $\mathrm{Re}\,\mu > 0$.

9. Neglecting rounding, and taking $x^{(0)} = 0$, show that for large enough values of k

$$\|x^{(k)} - x\| \leqq \frac{\varrho^k}{|1 - \lambda_1|} |\langle v^1, h \rangle| \cdot \|u^1\|$$

with $\varrho = |\lambda_1| \geqq |\lambda_2| \dots$, u^1 and v^1 are the eigenvector and reciprocal eigenvector of J; assuming J is diagonalizable and irreducible. Compare this bound and

$$\|x^{(k)} - x\| \le \gamma\alpha \max_i \left| \frac{1}{1 - \lambda_i} \right|$$

where $\alpha > \|x^{(k+1)} - x^{(k)}\|$

10. If in Jacobi's method,

$$\|\hat{x}^{(k)} - x\| \le \gamma\alpha \max_i \left| \frac{1}{1 - \lambda_i} \right| + 0(\|\Delta J\|, \|\Delta h\|)$$

obtain an expression for $0(\|\Delta J\|, \|\Delta h\|)$. Take $\alpha \geqq \|\hat{x}^{(k+1)} - \hat{x}^{(k)}\|$.

11. If$B^{(0)}$ is an approximate inverse of A, show that $B^{(1)}$ equal to $B^0 + (A + \Delta A)^{-1} \times (I - AB^{(0)})$ — is a better solution. Write the steps in computing the inverse of A by the method of iterative refinement. Show that $B^k \to A^{-1}$ as k approaches infinity.

12. If $B^{(0)}$ is an approximate inverse of A, improve on $B^{(0)}$ using the method of iterative refinement

$$A = \begin{bmatrix} 3 & 2 & 1 \\ 2 & 2 \times 10^{-9} & 2 \times 10^{-9} \\ 1 & 2 \times 10^{-9} & -10^{-9} \end{bmatrix},$$

$$B^{(0)} = \begin{bmatrix} -7 \times 10^{-10} & 4.33 \times 10^{-1} & 1.33 \times 10^{-1} \\ 4.13 \times 10^{-1} & -1.03 \times 10^{8} & 2.06 \times 10^{8} \\ 1.72 \times 10^{-1} & 2.06 \times 10^{8} & -4.13 \times 10^{8} \end{bmatrix}$$

13. Given $AX^{(k+1)} = X^{(k)}B + C$ as the iterative form of Lyapunov's equation, show that the problem is equivalent to solving the equation $(I \otimes A - B^T \otimes I)\,x = c$ using the linear iteration $Mx^{(k+1)} = Nx^{(k)} + c$, and the splitting $M - N = I \otimes A - B^T \otimes I$. Prove that the process converges if

$$\varrho(M^{-1}N) = \varrho((I \otimes A)^{-1}(B^T \otimes I)) = \varrho(B^T \otimes A^{-1}) < 1\,,$$

i.e. the iterations will converge if

$$\max_i |\lambda_i(B)| < \min_i |\lambda_i(A)|$$

where λ stands for eigenvalue.

14. If C is an approximate inverse of A, and $B^{(k+1)} = B^{(k)} + C(I - AB^{(k)})$ show that $B^{(k)}$ approaches A^{-1} slower than in the case where we use the expression $B^{(k+1)} = B^{(k)} + B^{(k)}(I - AB^{(k)})$.

15. Obtain, using an interval iterative scheme, values of x^I, y^I and z^I that enclose the solution of $A^I x = b^I$ where

$$A^I = \begin{bmatrix} [2 \pm 0.1] & [1 \pm 0.1] \\ [1 \pm 0.1] & [1 \pm 0.1] \end{bmatrix}, \qquad b^I = \begin{bmatrix} [3 \pm 0.1] \\ [2 \pm 0.1] \end{bmatrix}$$

16. Show that $X_{k+1}^I = \{(I - BA^I)\, X_k^I + B\} \cap X_k^I$ improves upon an inverse of A Show how to choose B and X_0.

17. If B is an approximate inverse of A^I, show that $B_{k+1}^I = B + (I - BA^I)\, B_k^I$ approximates $(A^I)^{-1}$ for large k. Show that a suitable B_0 can be chosen by

$$B_0 = [-1,1] \frac{\|B\|}{1 - \|I - BA^I\|}, \qquad \forall i, j$$

(see also Thieler (1975) for other iterative schemes).

18. If $Ax = b$ has an approximate solution $\hat{x}$ with residual r, show that

$$\|\hat{x} - x\| \leqq \frac{\|Br\|}{1 - \|BA - I\|}$$

where B is the approximate inverse of A related to the solution.

19. Roughly speaking, if the residual $r = Ax^{(i)} - b$ is computed with a precision $\varepsilon/\text{cond}\,(A)$, show that the method of iterative refinement will produce a solution $\hat{x}$ that is correct to full precision.

20. Show that the method of iterative refinement for updating an initial approximation $x^{(0)}$ to $Ax = b$ will produce a full precision solution whenever ε cond $(A) < 1$.

Chapter 4

The Least-Squares Problem

4.1 Introduction

The set of linear simultaneous equations $Ax = b$, $A \in \mathbb{R}^{m \times n}$, has either a unique solution for x, more than one solution for x or no solution at all. For x to be unique, A is necessarily nonsingular and x is expressed as $x = A^{-1}b$ $(m = n)$. The situation in which there is more than one solution occurs when b can be expressed linearly in some few column vectors of A having rank equal to $r(A) < n$. The equations are said to be consistent yet indeterminate. The solution comes as $x = A^i b$, where A^i is some generalized inverse of A satisfying $AA^iA = A$. A generalized inverse A^i satisfying the latter condition can be easily suggested (Bellman 1970, p. 105) as

$$A^i = P \begin{bmatrix} I_r & X \\ Y & Z \end{bmatrix} R$$

where R and P are respectively elementary row and column operations which bring A to the canonical form, namely

$$RAP = \begin{bmatrix} I_r & 0 \\ 0 & 0 \end{bmatrix}$$

while X, Y and Z are arbitrary. x is therefore given by

$$x = A^t b + (I - A^t A)\, c$$

where A^t is the determinate part of A^i and c is an arbitrary vector accounting for X, Y and Z. To see how this equivalence follows we note that $A(I - A^t A) = A(I - A^i A) = 0$ and that $AA^t b = b$, so that premultiplying x by A yields b on the right hand side as our starting assumption. But the vectors of $(I - A^t A)$ lie in the null space of A, and a linear combination of them would represent the homogeneous solution. As to $AA^t b$ being equal to b, this holds true if the equations are consistent i.e. $r(A) = r(A \mid b)$ or that Rb has the last $m - r$ elements made equal to zero from consistency, where A is of dimension $m \times n$. For in this case

$$Rb = \operatorname{diag}(I_r \mid 0)\, Rb$$

or that

$$b = R^{-1} \begin{bmatrix} I_r & 0 \\ 0 & 0 \end{bmatrix} \quad P^{-1} P \begin{bmatrix} I_r & 0 \\ 0 & 0 \end{bmatrix} Rb = AA^t b$$

Although A^i is obtained systematically, it is not unique. Other forms of A^i exist which satisfy additional conditions. One might further impose $A^iAA^i = A^i$, to imply $Z = YX$. A^i is then called a *reflexive generalized inverse*. Furthermore, if we impose $(AA^i)^* = AA^i$ and $(A^iA)^* = A^iA$, we are led to more restrictions on X, Y, R and P. Indeed we can do this, since R and P are not unique and X and Y are arbitrary. The last two conditions imply that $X = 0$, $Y = 0$ and that each of the matrices R and P is the product of a unitary matrix and a diagonal one; leading directly to the *singular value decomposition* of A, namely

$$U^*AV = \begin{bmatrix} D & 0 \\ 0 & 0 \end{bmatrix}, \qquad D = \operatorname{diag}(\sigma_1, \dots, \sigma_r)$$

and

$$A^i = V \begin{bmatrix} D^{-1} & 0 \\ 0 & 0 \end{bmatrix} U^*,$$

where U and V are two unitary matrices related to A as follows:

$$U^*AA^*U = \operatorname{diag}(\sigma_1^2, \dots, \sigma_r^2, 0, \dots, 0)_{m \times m}$$

and

$$V^*A^*AV = \operatorname{diag}(\sigma_1^2, \dots, \sigma_r^2, 0, \dots, 0)_{n \times n}$$

$\sigma_1, \dots, \sigma_r$ are called the singular values of A, i.e. $\sigma_i = \sqrt{\lambda_i(A^*A)} = \sqrt{\lambda_i(AA^*)}$, where λ stands for a nonzero eigenvalue. The above relationships now bring about a systematic computation of A^i. Such an A^i, which now satisfies $AA^iA = A$, $A^iAA^i = A^i$, $(A^iA)^* = A^iA$ and $(AA^i)^* = AA^i$, is uniquely determined and is called the *Moore-Penrose inverse* defined for an arbitrary $m \times n$ matrix A and denoted A^+. The solution x would be given by

$$x = A^+b + (I - A^+A)\,c$$

This generalized inverse was introduced by Moore in 1920, but rediscovered by Penrose in 1955. For background on generalized inverses, the reader should consult the texts by Rao and Mitra (1971), Bouillon and Odell (1971) and Ben-Israël and Greville (1974).

Finally, the situation in which the equations $Ax = b$ admit of no solution occurs when the latter are inconsistent, i.e. $r(A) \neq r(A : b)$. We are then contented with a least-squares fitting or a solution which minimizes the residual error $e = Ax - b$ in the least-squares sense. Such a situation is often encountered in physical experiments, where the number of observations exceeds that of the unknowns (*overdetermined system*). For although n measurements should be enough to determine the n unknowns, we usually take many more since we lack accuracy in each observation. We then end up with an odd situation: n measurements are enough to determine the unknowns, but the unknowns computed do not satisfy the rest of the measurements. To alleviate this difficulty, Gauss invented the least-squares technique, by which a straight line $y = ax + b$ can be made to fit three or more non-collinear points. If $m > 2$ is the number of such points (x_i, y_i), then the problem is set out so as to find a and b such that

$$F = \sum_{i=1}^{m} (ax_i + b - y_i)^2$$

is minimized. Each bracket constitutes the difference between each of the expected and observed values of the dependent variable y. The values of a and b are then obtained from

$$\frac{\partial F}{\partial a} = 0, \qquad \frac{\partial F}{\partial b} = 0$$

yielding the two linear equations

$$\begin{bmatrix} m & \sum_{i=1}^{m} x_i \\ \sum_{i=1}^{m} x_i & \sum_{i=1}^{m} x_i^2 \end{bmatrix} \begin{bmatrix} b \\ a \end{bmatrix} = \begin{bmatrix} \sum_{i=1}^{m} y_i \\ \sum_{i=1}^{m} x_i y_i \end{bmatrix}.$$

Another suggestion for F is

$$F = \frac{1}{1 + a^2} \sum_{i=1}^{m} (ax_i + b - y_i)^2,$$

see Fröberg (1969) and Usmani and Chebib (1978). Such a choice for F allows for minimizing the sum of the squares of perpendicular distances drawn from (x_i, y_i) to the line $y = ax + b$. However the first choice for F is the most popular for reasons outlined later.

Likewise, for $Ax = b$ (A of dimension $m \times n$, $m > n$), x may be found by minimizing

$$F = \sum_{i=1}^{m} \left| \sum_{j=1}^{n} a_{ij} x_j - b_i \right|^p, \qquad p \geqq 1.$$

A common value of p is $p = 2$ and we reach the least-squares fitting. Setting $p = 2$ and $\frac{\partial F}{\partial x_j} = 0$, $j = 1, \dots, n$, one can easily obtain

$$A^*Ax = A^*b$$

to be solved for a unique x, if $r(A) = n$. If $r(A) < n$, A^*A is singular, but the above equations are consistent, though $Ax = b$ is not; for to propose that $x = A^i b$ is a solution for some generalized inverse A^i, suggests that $A^*AA^i b = A^*b$, $\forall b$, even if $AA^ib \neq b$. The equations $A^*Ax = A^*b$ are therefore consistent, if $A^*AA^i = A^*$, or equivalently $(AA^i)^* A = A$. Hence $x = A^ib$ is a solution to $A^*Ax = A^*b$ whenever A^i satisfies

$$(AA^i)^* = AA^i, \qquad AA^iA = A$$

A generalized inverse satisfying the above two restrictions only, is called a *pseudo-inverse* of A. Such an inverse is quite enough to solve the least-squares problem as was first observed by Rao in 1955. However for the sake of generalization, we can still work out our problem using A^+, and show that $x = A^+b + (I - A^+A)\,c$ is a solution to $A^*Ax = A^*b$. Indeed it is, since the axioms of the pseudo-inverse are satisfied. For x to be a solution, one must habe $A^*AA^+b = A^*b$ and $A^*A(I - A^+A) = 0$ being satisfied as above, though x written in this form serves as a generalization to the solution of the many classes of $Ax = b$. Notice the elegance in writing $x = A^+b$ as a substitute for $x = (A^*A)^{-1} A^*b$, when $r(A) = n$. We therefore conclude that $x = A^+b + (I - A^+A)\,c$ is the general least-squares solution to $Ax = b$. Only when the equations are inconsistent, does x become the vector which minimizes the error between Ax and b in the least-squares sense (in the consistent case, the residual vector comes as $(I - AA^+)\,b = 0$). Another advantage for using A^+ instead of being confined to an A^i with $AA^iA = A$ and $(AA^i)^* = AA^i$ is that one may seek a *minimum norm solution*. In this case A^i must further satisfy $A^iAA^i = A^i$ and $(A^iA)^* = A^iA$ (see exercise 4.2). $x = A^+b$ is therefore called a *minimal least-squares solution* of $Ax = b$.

4.2 Perturbations of the Moore-Penrose Inverse

Unlike the simple situation in which $\det(A) \neq 0$, the solution $x(\varepsilon)$ of $(A + \varepsilon A_1) \times x(\varepsilon) = b + \varepsilon b_1$, when A is singular, is more complicated to obtain in a series of ε. For although $x(\varepsilon)$, when A is nonsingular, can always be represented in the form (cf. Sect. 1.4)

$$x(\varepsilon) = x + \varepsilon x_{(1)} + \varepsilon^2 x_{(2)} + \ldots$$

where $x = A^{-1}b$, this does not generally apply to the case where A is singular or rectangular. The reason is that although $(A + \varepsilon A_1)^{-1}$ may exist, its expansion into a Taylor's series in ε is just impossible; for when $\varepsilon = 0$, A^{-1} does not exist. The case when $\det(A + \varepsilon A_1)$ vanishes identically in ε, or when A is rectangular, is even more involved since only a generalized inverse of $(A + \varepsilon A_1)$ can be evaluated.

Such a problem appears either in indeterminate systems, where $Ax = b$ is a consistent set of eqúations having $x = A^+b + (I - A^+A)\,c$ as a solution, or in overdetermined systems in which x minimizes the residual $e = Ax - b$ in a least-squares sense. For when A and b both undergo a slight perturbation $A + \varepsilon A_1$ and $b + \varepsilon b_1$, the solution $x(\varepsilon)$ of the perturbed system, in a series of ε, will have to be evaluated through an expansion of $(A + \varepsilon A_1)^+$, around $\varepsilon = 0$. We shall find that

$x(\varepsilon)$, $(A + \varepsilon A_1)^+$ or $(A + \varepsilon A_1)^{-1}$ will all assume in general a Laurent's expansion rather than a Taylor's one. To show this on some examples consider

$$\begin{bmatrix} 1 + \varepsilon & -1 \\ 2 & -2 + 2\varepsilon \end{bmatrix} \begin{bmatrix} x_1 \\ x_2 \end{bmatrix} = \begin{bmatrix} 1 \\ 2 \end{bmatrix}$$

having as solutions $x_1 = x_2 = 1/\varepsilon$. And

$$\begin{bmatrix} 1 + \varepsilon & -1 \\ 2 & -2 + 2\varepsilon \end{bmatrix}^{-1} = \frac{1}{\varepsilon^2} \begin{bmatrix} -1 & 1/2 \\ -1 & 1/2 \end{bmatrix} + \frac{1}{\varepsilon} \begin{bmatrix} 1 & 0 \\ 0 & 1/2 \end{bmatrix}$$

also

$$\begin{bmatrix} 1 & -1 + \varepsilon & -2 \\ 3 & -3 & -6 - 2\varepsilon \end{bmatrix}^+ = \frac{1}{29\varepsilon + 16\varepsilon^2 + 4\varepsilon^3} \begin{bmatrix} 21 + 4\varepsilon & -7 + 3\varepsilon \\ 33 + 20\varepsilon + 4\varepsilon^2 & -11 - 4\varepsilon \\ -6 - 6\varepsilon & 2 - 2\varepsilon - 2\varepsilon^2 \end{bmatrix}$$

One feature therefore of $(A + \varepsilon A_1)^+$ is that, in general

$$\lim_{\varepsilon \to 0} (A + \varepsilon A_1)^+ \neq A^+ ;$$

i.e. $(A + \varepsilon A_1)^+$ is not continuous at $\varepsilon = 0$. Only when $(A + \varepsilon A_1)^+$ assumes a Taylor's expansion around $\varepsilon = 0$ does such a limit exist. A perturbation which guarantees the above limit is said to be in the *acute case* after Wedin (1973) or *acute perturbation* according to Stewart (1977). But in general $(A + \varepsilon A_1)^+$ takes the form

$$(A + \varepsilon A_1)^+ = \sum_{i=1}^{s} \frac{A_{-i}}{\varepsilon^i} + \sum_{i=0}^{\infty} \varepsilon^i A_{+i}$$

To obtain the A_{-i} and A_{+i} matrices, Deif (1983b) suggested a singular value decomposition of $(A + \varepsilon A_1)$. Let A be an $m \times n$ matrix, with $m \leqq n$ without loss of generality, and define the two unitary matrices V and U whose columns are eigenvectors of A^*A and AA^* respectively. Then if

$$U^*AV = \begin{bmatrix} D & 0 \\ 0 & 0 \end{bmatrix}$$

(see Sect. 4.1) with $D = \operatorname{diag}(\sigma_1, \ldots, \sigma_r)$, $r = \operatorname{rank}(A) = \operatorname{rank}(A^*A) = \operatorname{rank}(AA^*)$, and $\sigma_1, \ldots, \sigma_r$ are the nonvanishing singular values of A, i.e. $\sigma_k = \sqrt{\lambda_k(A^*A)} = \sqrt{\lambda_k(AA^*)}$, $k = 1, \ldots, r$, then,

$$A^+ = V \begin{bmatrix} D^{-1} & 0 \\ 0 & 0 \end{bmatrix} U^*$$

As for $(A + \varepsilon A_1)$, define similarly $\tilde{V}$, $\tilde{U}$ diagonalizing respectively through a similarity transformation $(A + \varepsilon A_1)^* (A + \varepsilon A_1)$ and $(A + \varepsilon A_1)(A + \varepsilon A_1)^*$. Let $\tilde{\sigma}_1, \ldots, \tilde{\sigma}_p$ $(r \leqq p \leqq m)$ be the nonvanishing singular values of $(A + \varepsilon A_1)$, then

$$\begin{aligned}\tilde{\sigma}_k &= \sqrt{\tilde{\lambda}_k(A^*A + \varepsilon(A^*A_1 + A_1^*A) + \varepsilon^2 A_1^* A_1)} \\ &= \sqrt{\tilde{\lambda}_k(AA^* + \varepsilon(AA_1^* + A_1A^*) + \varepsilon^2 A_1 A_1^*)}\,.\end{aligned}$$

and because A^*A and AA^* are both Hermitian with Hermitian perturbation, $\tilde{\lambda}_k$, $\tilde{v}^k$, an $\tilde{u}^k$ can be chosen such that they will assume an analytic expansion in the neighbourhood of $\varepsilon = 0$, namely,

$$\begin{aligned}\tilde{\lambda}_k &= \lambda_k + \varepsilon\lambda_k^{(1)} + \varepsilon^2\lambda_k^{(2)} + \ldots, & k &= 1, \ldots, p \\ \tilde{v}^k &= v^k + \varepsilon v_{(1)}^k + \varepsilon^2 v_{(2)}^k + \ldots, & k &= 1, \ldots, p \\ \tilde{u}^k &= u^k + \varepsilon u_{(1)}^k + \varepsilon^2 u_{(2)}^k + \ldots, & k &= 1, \ldots, p\end{aligned}$$

with $\lambda_k = 0$, $k = r + 1, \ldots, p$, so is $\lambda_k^{(1)}$, $k = r + 1, \ldots, p$. To show that $\lambda_k^{(1)} = 0$, $k = r + 1, \ldots, p$, we first note that $Av^k = 0$, $k = r + 1, \ldots, n$. The latter follows from $A^*Av^k = 0$, $k = r + 1, \ldots, n$ and hence

$$0 = A^{*+} A^* A v^k = (AA^+)^* Av^k = AA^+Av^k = Av^k\,, \qquad k = r + 1, \ldots, n\,,$$

and similarly $A^*u^k = 0$, $k = r + 1, \ldots, m$. One therefore has from Bellman (1970, p. 63)

$$\lambda_k^{(1)} = \langle v^k, (A^*A_1 + A_1^*A)\, v^k \rangle = 0$$

And it follows directly, for sufficiently small $|\varepsilon|$, that

$$\begin{aligned}(A + \varepsilon A_1)^+ &= \sum_{k=1}^{r} \frac{v^k\rangle\langle u^k + \varepsilon(v_{(1)}^k\rangle\langle u^k + v^k\rangle\langle u_{(1)}^k) + O(\varepsilon^2)}{\sqrt{\lambda_k + \varepsilon\lambda_k^{(1)} + \varepsilon^2\lambda_k^{(2)} + \ldots}} \\ &+ \sum_{k=r+1}^{p} \frac{v^k\rangle\langle u^k + \varepsilon(v_{(1)}^k\rangle\langle u^k + v^k\rangle\langle u_{(1)}^k) + O(\varepsilon^2)}{\sqrt{\varepsilon^2\lambda_k^{(2)} + \varepsilon^3\lambda_k^{(3)} + \ldots}}\,.\end{aligned}$$

The above expression for $(A + \varepsilon A_1)^+$ has one main disadvantage in that it requires first the explicit evaluation of $\lambda_k^{(i)}$, $v_{(i)}^k$ and $u_{(i)}^k$, $k = 1, \ldots, p$, $i = 1, 2, 3, \ldots$, which is usually tedious to obtain. It is therefore used to obtain $(A + \varepsilon A_1)^+$ up to a few terms only. But we shall use it to get some useful information about the expansion.

It now seems clear that $(A + \varepsilon A_1)^+$ will in general possess a laurent's expansion around $\varepsilon = 0$. The principal part of the series is obtained from the term

$$\sum_{k=r+1}^{p} \frac{v^k\rangle\langle u^k + \varepsilon(v_{(1)}^k\rangle\langle u^k + v^k\rangle\langle u_{(1)}^k) + O(\varepsilon^2)}{\sqrt{\varepsilon^2\lambda_k^{(2)} + \varepsilon^3\lambda_k^{(3)} + \ldots}}\,.$$

We immediately notice that, if $p = r$ i.e. $r(A + \varepsilon A_1) = r(A)$, this term vanishes, and we end up with a Taylor's expansion of $(A + \varepsilon A_1)^+$, for sufficiently small $|\varepsilon|$; i.e.

$$(A + \varepsilon A_1)^+ = \sum_{k=0}^{\infty} \varepsilon^k A_{+k}$$

In other words, acute perturbation is only obtained when $r(A)$ does not change under a slight perturbation εA_1. For example, consider

$$A = \begin{bmatrix} 1 & 0 \\ 0 & 0 \end{bmatrix}, \qquad A_1 = \begin{bmatrix} 0 & 1 \\ 0 & 0 \end{bmatrix}$$

we have

$$(A + \varepsilon A_1)^+ = \frac{1}{1 + \varepsilon^2} \begin{bmatrix} 1 & 0 \\ \varepsilon & 0 \end{bmatrix}.$$

The above result directly implies that if $\det(A) = 0$, then $(A + \varepsilon A_1)^{-1}$ once it exists, must possess a Laurent's expansion only. But what is the order of singularity of $(A + \varepsilon A_1)^+$, i.e. the value of the non-negative integer s denoting the order of the pole at $\varepsilon = 0$? This naturally depends on whether some of $\lambda_k^{(2)}$, $\lambda_k^{(3)}$, ... are zero. In other words if

$$\lambda_k^{(2)} = \lambda_k^{(3)} = 0\,, \qquad k = r + 1, \ldots, p \quad \text{and} \quad \lambda_k^{(4)} \neq 0$$

we have $s = 2$. Furthermore, if $\lambda_k^{(4)} = \lambda_k^{(5)} = 0$, $\lambda_k^{(6)} \neq 0$ then $s = 3$, and so on. It seems natural then to investigate the conditions under which $\lambda_k^{(i)} = 0, k = r + 1, \ldots, p$, $i = 1, 2, 3, \ldots$ It is found that a necessary and sufficient condition that $(A + \varepsilon A_1)^+$ has a multiple pole at $\varepsilon = 0$ of order s, is

$$Av_{(1)}^k + A_1 v^k = Av_{(2)}^k + A_1 v_{(1)}^k = \ldots = Av_{(s-1)}^k + A_1 v_{(s-2)}^k = 0, k = r + 1, \ldots, p\,.$$

For instance, if s were to be two, $\lambda_k^{(2,3)}$ must vanish implying that the first term above is zero. To prove such a proposition, write

$$(A + \varepsilon A_1)^* (A + \varepsilon A_1) (v^k + \varepsilon v_{(1)}^k + \ldots) = (\varepsilon^2 \lambda_k^{(2)} + \ldots) (v^k + \varepsilon v_{(1)}^k + \ldots)$$

and $\lambda_k^{(2)}$ being zero, gives upon grouping first and second order terms

$$A^*(Av_{(1)}^k + A_1 v^k) = 0$$

and

$$A^* A v_{(2)}^k + A^* A_1 v_{(1)}^k + A_1^*(Av_{(1)}^k + A_1 v^k) = 0\,.$$

The first equation directly suggests that

$$Av_{(1)}^k + A_1 v^k = \sum_{i=r+1}^{m} c_{ki} u^i$$

To show that $c_{ki} = 0$, substitute this last relationship into the second equation above. Taking the inner product with v^k yields

$$\langle v^k, A_1^* \Sigma c_{ki} u^i \rangle = 0 .$$

But we also have from the same relationship, that

$$\langle u^i, A_1 v^k \rangle = c_{ki} .$$

Hence $\Sigma |c_{ki}|^2 = 0$, giving $c_{ki} = 0$, and necessity is proved for $s = 2$. Sufficiency is easier; if

$$Av_{(1)}^k + A_1 v^k = 0 ,$$

then

$$\lambda_k^{(2)} = \langle v^k, A_1^*(Av_{(1)}^k + A_1 v^k) \rangle = 0 .$$

To show that $\lambda_k^{(3)} = 0$, we have

$$\lambda_k^{(3)} = \langle v^k, A_1^*(Av_{(2)}^k + A_1 v_{(1)}^k) \rangle .$$

But from the second equation above, one is left with

$$A^*(Av_{(2)}^k + A_1 v_{(1)}^k) = 0$$

i.e. that

$$Av_{(2)}^k + A_1 v_{(1)}^k = \Sigma \beta_{ki} u^i$$

and

$$\lambda_k^{(3)} = \Sigma \beta_{ki} \langle v^k, A_1^* u^i \rangle = 0 .$$

To ensure the above conditions for the order of the pole at $\varepsilon = 0$ of $(A + \varepsilon A_1)^+$ looks quite complicated. However for the special case in which A and A_1 are square matrices and commute, one has simply that $s = \nu =$ index A. By ν we denote the least non-negative integer for which $r(A^\nu) = r(A^{\nu+1})$. For example, for diagonalizable matrices $\nu = 1$, and for a nonsingular matrix A, $\nu = 0$. It also stands for the size of the largest Jordan block corresponding to the zero eigenvalue of A. To show that $s = \nu$ for commutative matrices A and A_1, it is enough to prove that, for instance, for $\nu = 2$,

$$Av_{(1)}^k + A_1 v^k = 0 , \quad \text{for some } k \quad (r + 1 \leqq k \leqq p) ,$$

while

$$Av_{(2)}^k + A_1 v_{(1)}^k \neq 0 .$$

Taking $k = r + 1$ and observing that

$$Av_{(1)}^{r+1} + A_1 v^{r+1} = \sum c_{r+1,i} u^i$$

and $A_1 v^{r+1} = \mu v^{r+1}$ (A and A_1 commute), then by taking the inner product with u^{r+1} yields $c_{r+1,i} = 0, \forall i$. This also shows that v^{r+1} is proportional to the second vector in the chain of v vectors. Let this vector be $\tilde{v}$, then $A_1 v_{(1)}^{r+1} \sim \tilde{v}$ and

$$Av_{(2)}^{r+1} + A_1 v_{(1)}^{r+1} = Av_{(2)}^{r+1} + \beta\tilde{v} = \sum_i \alpha_{r+1,i} u^i .$$

But because

$$\langle u^{r+1}, \tilde{v}\rangle \neq 0 \quad \text{for} \quad v = 2 ,$$

one has

$$Av_{(2)}^{r+1} + A_1 v_{(1)}^{r+1} \neq 0 ;$$

thus completing the proof. In fact, this special result was first discovered by Rose (1978). He proved that if A and A_1 commute and $(A + \varepsilon A_1)^{-1}$ exists, then it can be written explicitly in the form

$$(A + \varepsilon A_1)^{-1} = A^D \sum_{k=0}^{\infty} (-)^k (A^D A_1)^k \varepsilon^k + A_1^D (I - AA^D) \sum_{k=0}^{v-1} (-)^k (AA_1^D)^k \varepsilon^{-k-1} .$$

A^D is the *Drazin inverse* of A and is the unique solution to $AA^D = A^D A$, $A^D A A^D = A^D$ and $A^{k+1} A^D = A^k$, $k \geqq v$. For more detailed information about the Drazin inverse and related applications, the reader is referred to Campbell and Meyer (1979).

Now returning to the interesting case of acute perturbation, i.e. when $r(A + \varepsilon A_1) = r(A)$, we have said that $(A + \varepsilon A_1)^+$ will assume a Taylor's series around $\varepsilon = 0$, that is

$$(A + \varepsilon A_1)^+ = \sum_{k=0}^{\infty} \varepsilon^k A_{+k} .$$

Our next task is to determine A_{+k}. It was Ben-Israël (1966) who first derived this expression without explicity stating that $r(A + \varepsilon A_1) = r(A)$, yet in fact he asserted this condition by two other conditions. Ben-Israël showed that if $AA^+A_1 = A_1$, $A^+AA_1^* = A_1^*$ and $\|\varepsilon A^+ A_1\| < 1$, then

$$(A + \varepsilon A_1)^+ = A^+ + \sum_{k=1}^{\infty} (-)^k (A^+A_1)^k A^+ \varepsilon^k .$$

His proof relies upon the identity $(A + \varepsilon A_1) = A(I + \varepsilon A^+ A_1)$. Also by denoting $C = (I + \varepsilon A^+ A_1)$, the author showed that

$$A^+ ACC^* A^* = CC^* A^* .$$

But we also have that $CC^+ A^* AC = A^* AC$ as C is nonsingular from $\|\varepsilon A^+ A_1\| < 1$. It follows from exercise 4.1 that

$$(A + \varepsilon A_1)^+ = (I + \varepsilon A^+ A_1)^{-1} A^+ ,$$

from which the above expansion is directly obtained. The above series in ε becomes a simpler formula to use than,

$$(A + \varepsilon A_1)^+ = \sum_{k=1}^{r} \frac{v^k\rangle\langle u^k + \varepsilon(v^k_{(1)}\rangle\langle u^k + v^k\rangle\langle u^k_{(1)}) + O(\varepsilon^2)}{\sqrt{\lambda_k + \varepsilon\lambda_k^{(1)} + \varepsilon^2\lambda_k^{(2)} + \ldots}}$$

provided that A_1 satisfies the Ben-Israël conditions. Note that the principal part of $(A + \varepsilon A_1)^+$ vanishes and the above analytic part remains only. The reason is that $r(A) = r(A + \varepsilon A_1)$; a result which follows immediately from using two only of the Ben-Israël three conditions (see exercise 4.15). Nevertheless, his above expansion of $(A + \varepsilon A_1)^+$ into a series of ε, was established using all three conditions.

It now becomes apparent that unless $r(A + \Delta A) = r(A)$, $(A + \Delta A)^+ - A^+$ is always large. In other words, for $r(A + \Delta A) \neq r(A)$, although $\|\Delta A\|$ can be made sufficiently small, $\|(A + \Delta A)^+ - A^+\|$ will become unbounded. Since it is interesting to obtain finite bounds on $\|(A + \Delta A)^+ - A^+\|$ as $\Delta A \to 0$, we shall confine ourselves to acute perturbation.

For the special case in which $AA^+ \Delta A = \Delta A$, $A^+ A \, \Delta A^* = \Delta A^*$ and $\|A^+ \Delta A\| < 1$, one easily obtains from $(A + \Delta A)^+ - A^+ = [(I + A^+ \Delta A)^{-1} - I] A^+$, that

$$\frac{\|(A + \Delta A)^+ - A^+\|}{\|A^+\|} \leqq \frac{\|A^+ \Delta A\|}{1 - \|A^+ \Delta A\|}$$

(cf. exercise 1.18). If $\|A^+ \Delta A\| < 1$ is further replaced by $\|A^+\| \, \|\Delta A\| < 1$, we reach a bound similar to that of Sect. 1.4 (cf. exercise 4.14). The condition number to the problem of computing A^+ becomes $\|A\| \, \|A^+\|$, generalizing cond (A) to include the case where A is singular. Also, similar to the condition number of a nonsingular matrix measuring its distance to the closest singular matrix (see Sect. 1.3), such a more general condition number here measures the closeness of A to one of smaller rank (see Demko (1985)).

The above bound, as first derived by Ben-Israël (1966), was one of the earliest bounds for perturbed pseudo-inverses. Unfortunately it suffers from one major drawback: regarding the conditions imposed on ΔA. The latter is usually uncontrollable and cannot adhere to certain rules. But acute perturbations do not imply rules as such; the condition $\|A^+\| \, \|\Delta A\| < 1$ is usually guaranteed under slight perturbation ΔA. As for $r(A + \Delta A) = r(A)$, it is only invoked to ensure a finite bound for $\|(A + \Delta A)^+ - A^+\|$ as ΔA approaches zero. Truly the Ben-Israël conditions imply acute perturbations, but the opposite is not necessarily true (see exercise 4.15).

Bounds for acute perturbations were first deduced by Wedin (1973). He showed that

$$\|(A + \Delta A)^+\|_2 \leqq \frac{\|A^+\|_2}{1 - \|A^+\|_2 \|\Delta A\|_2},$$

a result which is very similar to an anlogous one obtained previously for nonsingular perturbations (see exercise 1.18). To reach the above bound, Wedin used an interesting inequality relating the singular values of both A and $(A + \Delta A)$. By substituting in exercise 4.5,

$$\sigma_r(A + \Delta A) = 1/\|(A + \Delta A)^+\|_2 \quad \text{and} \quad \sigma_r(A) = 1/\|A^+\|_2,$$

yields directly the above result. But the main contribution of the Wedin paper lies in providing a simple decomposition theorem for any ΔA. Wedin showed that (see exercise 4.17)

$$B^+ - A^+ = -B^+ \Delta A A^+ + B^+(I - AA^+) - (I - B^+B) A^+,$$

where $B = A + \Delta A$. The above relation was made more useful when the author noticed from $A^*(I - AA^+) = 0$, that

$$B^+(I - AA^+) = B^+ B^{+*} B^*(I - AA^+) = B^+ B^{+*} \Delta A^*(I - AA^+).$$

Similarly by dealing with the last term, one has that,

$$B^+ - A^+ = -B^+ \Delta A A^+ + B^+ B^{+*} \Delta A^*(I - AA^+) + (I - B^+ B) \Delta A^* A^{+*} A^+.$$

The above formula hence provides a bound for $\|B^+ - A^+\|_2$ for any A and B, that is

$$\|B^+ - A^+\|_2 \leqq (\|A^+\|_2 \|B^+\|_2 + \|B^+\|_2^2 + \|A^+\|_2^2) \|\Delta A\|_2.$$

More briefly

$$\|B^+ - A^+\| \leqq \mu \max \{\|A^+\|_2^2, \|B^+\|_2^2\} \|\Delta A\|_2$$

where $\mu = (1 + \sqrt{5})/2$ for the spectral norm, $\sqrt{2}$ for the Frobenius norm and 3 for an arbitrary norm (see Stewart 1977). Also for $r(A + \Delta A) = r(A)$ — being an important special case — one obtains

$$\frac{\|(A + \Delta A)^+ - A^+\|}{\|(A + \Delta A)^+\|_2} \leqq \mu \|A^+\|_2 \|\Delta A\|_2$$

where μ is given in the following table provided by Wedin (1973)

rank $\|\cdot\|$	Arbitrary	Spectral	Frobenius
$r(A) < \min(m, n)$	3	$(1 + \sqrt{5})/2$	$\sqrt{2}$
$r(A) = \min(m, n)$ $m \neq n$	2	$\sqrt{2}$	1
$r(A) = m = n$	1	1	1

The reader should note that, in the above bound, the condition $\|A^+\| \, \|\Delta A\| < 1$ is released. In fact, much earlier than that, Forsythe and Moler (1967) made a similar statement concerning nonsingular perturbations (see Sect. 1.4). This leads us to the heart of the matter; that of deriving an a-priori bound for acute perturbations similar to the one described in Sect. 1.4. By using the bound for $\|(A + \Delta A)^+\|$ provided previously, on finally reaches:

$$\frac{\|(A + \Delta A)^+ - A^+\|}{\|A^+\|_2} \leq \mu \frac{\operatorname{cond}_2(A) \dfrac{\|\Delta A\|_2}{\|A\|_2}}{1 - \operatorname{cond}_2(A) \dfrac{\|\Delta A\|_2}{\|A\|_2}} \, .$$

where $\operatorname{cond}(A) = \|A\|_2 \, \|A^+\|_2$ is the *spectral pseudo-condition number* measuring the sensitivity of A^+ to acute perturbations in A.

To see how the above bound works, suppose we want to predict the error in computing A^+ of a matrix A of size $m \times n$, when performed on an t-digit machine. Naturally, we shall limit ourselves to acute perturbations, this being the restriction imposed for the validity of our bound. In other words, we shall take it for granted that rank A does not change while computing A^+. To illustrate this on an example, we depict one already worked out by Lawson and Hanson (1974, p. 14) using the QU decomposition. One can obtain approximately for,

$$A = \begin{bmatrix} 0.4087 & 0.1594 \\ 0.4302 & 0.3516 \\ 0.6246 & 0.3384 \end{bmatrix}$$

that

$$B^+ = \begin{bmatrix} 3.1317 & -2.9327 & 1.5717 \\ -4.4785 & 6.0943 & -1.2669 \end{bmatrix}$$

on a five-digit machine, and by applying the above results, one gets:

$$\begin{aligned} \frac{\|B^+ - A^+\|_2}{\|B^+\|_2} &\leqq \sqrt{2}\,(\sigma_1(A)/\sigma_r(A))\, 5 \times 10^{-5} \sqrt{m \times n} \\ &\cong 1.5 \times 10^{-3} \end{aligned}$$

To check the goodness of the above bound, we computed A^+ exactly using a refinement technique (see Sect. 4.3) and substracted it from B^+, obtaining approximately

$$\Delta A^+ \cong \begin{bmatrix} 1.0861 \times 10^{-3} & 5.788 \times 10^{-5} & -8.1129 \times 10^{-4} \\ -1.922 \times 10^{-3} & -7.535 \times 10^{-5} & 1.4683 \times 10^{-3} \end{bmatrix}.$$

Hence the exact accuracy is governed by:

$$\frac{\|\Delta A^+\|_2}{\|B^+\|_2} \cong 3 \times 10^{-4} \, ,$$

being naturally somewhat better.

What follows next is an estimation of a bound for $\|\Delta x\|/\|x\|$, where x minimizes $\|Ax - b\|_2$ and $x + \Delta x$ minimizes $\|(A + \Delta A)(x + \Delta x) - (b + \Delta b)\|_2$. One has from $x + \Delta x = (A + \Delta A)^+ (b + \Delta b)$ that

$$\Delta x = [(A + \Delta A)^+ - A^+]\, b + (A + \Delta A)^+ \Delta b\,.$$

Then by using Wedin's decomposition theorem for $B^+ - A^+$, one obtains,

$$\Delta x = -B^+ \Delta A x + B^+ B^{+*} \Delta A^* r + (I - B^+ B) \Delta A^* A^{+*} x + B^+ \Delta b\,.$$

It follows upon taking norms, that

$$\frac{\|\Delta x\|_2}{\|x\|_2} \leqq \|B^+\|_2 \|\Delta A\|_2 + \|B^+\|_2^2 \|\Delta A\|_2 \frac{\|r\|_2}{\|A\|_2 \|x\|_2} \|A\|_2 + \|\Delta A\|_2 \|A^+\|_2$$

$$+ \|B^+\|_2 \frac{\|\Delta b\|_2}{\|b\|_2} \frac{\|b\|_2}{\|A\|_2 \|x\|_2} \|A\|_2$$

and by setting,

$$\alpha = \frac{\|\Delta A\|}{\|A\|}, \qquad \beta = \frac{\|\Delta b\|}{\|b\|}, \qquad \gamma = \frac{\|b\|}{\|A\| \|x\|}, \qquad \varrho = \frac{\|r\|}{\|A\| \|x\|}$$

$$k = \operatorname{cond}(A) = \|A\| \|A^+\|, \qquad \hat{k} = \frac{k}{1 - k\alpha}$$

according to Lawson and Hanson (1974), one ends up with the following bound:

$$\frac{\|\Delta x\|_2}{\|x\|_2} \leqq \hat{k}[(2 + \hat{k}\varrho)\,\alpha + \beta\gamma]\,.$$

This bound dominates $\|\Delta x\|/\|x\|$ for acute perturbations. Some slightly better looking expressions exist for the full rank case. The reader can check easily that, for $r(A) = n < m$, the above expression becomes $\hat{k}[(1 + \hat{k}\varrho)\,\alpha + \beta\gamma]$. For $r(A) = m < n$, it reads $\hat{k}(2\alpha + \beta)$. Finally for $r(A) = m = n$, it reduces to the simple expression $\hat{k}(\alpha + \beta)$ (cf. Sect. 1.4).

Perturbation theory of pseudo-inverses is the result of the work of many authors. Golub and Wilkinson (1966) were probably the first to notice the effect of the factor $\|A^+\| \|A\|$ on the solution of the least-squares problem. The problem of extending perturbation theory from inverses to pseudo-inverses was first studied by Ben-Israël (1966) under some strong restrictions on ΔA. Perhaps Wedin (1973) was the first to give the most general survey on the general perturbation theory of pseudo-inverses. A good review of the subject is also to be found in Stewart (1977). The term *acute perturbation* originated in Wedin (1973), but the expansion of $(A + \varepsilon A_1)^+$ into a Taylor's series in ε appeared in a different context in Stewart (1969). He showed that, under a sequence of matrices T_n satisfying $\|A^+\| \|T_n\| < 1$ and $\lim_{n\to\infty} T_n = 0$,

$$\lim_{n\to\infty} (A + T_n)^+ = A^+$$

if $r(A + T_n) = r(A)$. In fact, this property was discovered much earlier by Penrose (1955) who stated that A^+ is a continuous function of A if $r(A)$ is kept fixed.

But if $r(A + \Delta A) > r(A)$, the computation of A^+ is at stake, for $\|(A + \Delta A)^+\| \geqq 1/\|\Delta A\|$; a result which had first appeared in Stewart (1969). As to the expansion of $(A + \varepsilon A_1)^+$ or $(A + \varepsilon A_1)^{-1}$ into a Laurent's series, we mention Ben-Israël (1966), Rose (1978) and the author (1983 b). But the first considered only a class of acute perturbations, the second was only interested in $(A + \varepsilon A_1)^{-1}$ whereas to the third, the expansion relies upon shifts in the eigenvectors and can be useful for calculating only a few terms. For a Taylor's series expansion of functions of linear operators, the reader may consult Senechal (1985). Langenhop (1971, 1973) investigated the conditions under which $(A + \varepsilon A_1)^{-1}$ exists. For properties of $A(z)^{-1}$ of an analytic linear operator valued function $A(z)$, the reader may refer to Ribaric and Vidav (1969). Approximations of generalized inverses of linear operators can be found in Moore and Nashed (1974) and also Nashed (1976). Many results pertaining to perturbation theory of generalized inverses are set up in terms of perturbation of the singular values. This can be found in Thompson (1976), Wedin (1972), Stewart (1979) and Sun (1983). Perturbation theory for the least-squares problem with linear constraints is studied in Elden (1980).

4.3 Accuracy of Computation

Suppose we attempt to compute A^+ of a full rank matrix A using the most straightforward procedure

$$A^+ = (A^*A)^{-1} A^* ,$$

by solving the equations $(A^*A)\,X = A^*$. Though A^*A is positive definite and no pivoting strategy is required, X usually fails to be an accurate representation of A^+. The reason is that $\operatorname{cond}_2 (A^*A) = \operatorname{cond}_2^2 (A)$ (see exercise 4.19), and a large value of cond (A) will make cond (A^*A) much larger, indicating that the equations $(A^*A)\,X = A^*$ are generally ill-conditioned. This has already been seen in Sect. 4.2. For although the error in computing A^+ is at best governed by cond (A), that of solving the least-squares problem is affected by $\operatorname{cond}^2 (A)$. Stoer and Bulirsch (1980) asserted this same accuracy by applying the perturbation analysis of Sect. 1.4 on the equations $A^*Ax = A^*b$ to obtain similar bounds to those of Sect. 4.2 (see execises 4.20, 21). This gives the reader an idea about the futility of solving $A^*AX = A^*$ if it is to be used as a method for computing A^+.

To avoid this unnecessary worsening of the result, different methods have been proposed to compute A^+. Noble (1976), for instance, summarized some popular techniques. A famous one for computing A^+, when A is of full rank, is the QU decomposition. A can be decomposed using Householder transformations or the modified Gram-Schmidt process, into the form

$$A = QU$$

where Q has orthonormal columns and U is upper triangular. And it follows rather easily that

$$A^+ = U^{-1}Q^*$$

This method is stable by virtue of the backward stability of Householder transformations (see Sect. 1.5). The accuracy of A^+ is governed by cond $(A) =$ cond (U), being the factor governing the accuracy of U^{-1}. For a perturbation analysis of the QU decomposition, the reader is referred to Stewart (1977). The author provided bounds on the perturbations in both Q and U while preserving the same decomposition This technique has been widely implemented and tested many times in different contexts. For the numerical aspects of the method and Fortran programs, the reader should consult Lawson and Hanson (1974). Businger and Golub (1965) also studied it and their method is discussed in Wilkinson and Reinsch (1971) among a collection of available Algol procedures in numerical linear algebra. It is also included in some up-to-date packages like IMSL, NAG, EISPACK and LINPACK.

Unlike the simple case in which A has full rank, the situation in which A has a rank $< \min(m, n)$ is more vulnerable. The reason is that in the above case ΔA constitutes an acute perturbation for any small $\|\Delta A\|$, and the accuracy in computing A^+ is solely governed by cond (A). Unfortunately, this is not so in this case, where a perturbation ΔA generally amounts to an increase in $r(A)$ during computation. Though $\|\Delta A\|$ could be small, $(A + \Delta A)^+$ will indeed be large. Ironically enough, the smaller $\|\Delta A\|$ is, the more susceptible $(A + \Delta A)^+$ can be to perturbations. For we have from exercise 4.9 that $\|(A + \Delta A)^+\|_2 \geqq 1/\|\Delta A\|_2$. One cannot therefore proceed to compute A^+ without deciding about the rank of A. A very effective tool for determing $r(A)$ is the singular value decomposition. Knowing that $r(A) =$ *no.* of its nonzero singular values, one can therefore test these values against a small tolerance. And by setting the small singular values to zero, one effectively determines $r(A)$ (see Peters and Wilkinson (1970)). The same method can also be used to solve the least-squares problem, as was shown by Golub (1965). In fact, the singular value decomposition gives the most reliable determination of $r(A)$ according to Wilkinson in Wilkinson and Reinsch (1971, p. 7). Following this procedure, we impose a sort of acute perturbation, guaranteeing that $r(A + \Delta A) = r(A)$ for small $\|\Delta A\|$. The accuracy of computing A^+ is again governed by cond (A) and it is a remarkable technique.

At this stage, there should be a word of caution. Setting the small singular values to zero is not automatic, for they may be small already, thus comitting ourselves to an induced error. Even more serious, in some applications, the most interesting information is given by the small (but not negligibly small) singular values. Nash (1979, p. 35) provided a very interesting example to illustrate this awkward phenomenon. He computed the singular values of

$$A = \begin{bmatrix} 5 & 10^{-6} & 1 \\ 6 & 0.999999 & 1 \\ 7 & 2.00001 & 1 \\ 8 & 2.9999 & 1 \end{bmatrix}$$

using a twelve-digit machine. They are, up to six digits

$$\sigma_1 = 13.7530\,, \qquad \sigma_2 = 1.68961\,, \qquad \sigma_3 = 1.18853 \times 10^{-5}$$

and by determining the matrices U and V, he then worked out A_1^+ once ignoring σ_3 and, once, by taking it into account to yield A_2^+. He obtained two altogether different answers:

$$A_1^+ = \begin{bmatrix} 0.118521 & 0.070370 & 0.022219 & -0.025924 \\ -0.418538 & -0.170373 & 0.077795 & 0.325925 \\ 0.107415 & 0.048150 & -0.011116 & -0.070372 \end{bmatrix}$$

and

$$A_2^+ = \begin{bmatrix} 5322.32 & -1809.27 & -12349.4 & 8836.37 \\ -5322.78 & 1809.22 & 12349.9 & -8836.33 \\ -26610.8 & 9046.70 & 61747.0 & -44181.9 \end{bmatrix}$$

It looks as though the second answer is the correct one, simply because the author checked that $A_2^+ A_2$ approaches a unit matrix, while $A_1^+ A_1$ does not. By virtue of this dilema, Nash concluded that it is the user's responsibility to set up the tolerance criterion, i.e. to decide upon the degree of linear dependence in his data. The author went on making an elegant remark: "In a modelling situation for instance, the vectors of U and V corresponding usually to small singular values are almost certain to be largely determined by noise or errors in the original data. On the contrary, the same vectors when derived from tracking of a satellite may contain very significant information about orbit perturbations."

But this does not alter our understanding regarding the general concept of rank determination. Suppose the excessive singular values are zero, setting them to be non-zero will leave the matrix $(A^*A - \sigma_3^2 I)$ nonsingular, but computing the null vector of A^*A presupposes the latter also to be nonsingular (see Sect. 1.3). It is the machine's precision therefore which decides the tolerance and this must be the sole factor (assuming that no errors in A are present) in settling this critical issue.

However, in the least-squares problem, this situation is hardly dangerous; for the problem is not checking whether cond (A) is large or grows in computation only. Rather, it is mainly a problem of data fitting and hence any increase in the fitting variables, which makes the columns of A not really independent, increases in turn the possibility of finding small singular values which must be theoretically zero. Nash therefore suggested redefining the problem by leaving out certain columns of A from the best used to approximate b. In the *LLSQF* and *LSVDF* Fortran subroutines of the *IMSL* package, implementing Lawson and Hansons' programs (1974), a tolerance value is set so as to control the ratio between the smallest and largest singular values. It also determines the number of columns of A to be included in the basis for the least-squares fit of b. The process terminates when the inclusion of the next column would result in a matrix with condition number $\geqq 1/\text{tol}$. The reader is also advised to consult Golub and Kahan (1965) and Golub and Reinsch (1970) on numerical aspects of the singular value decomposition. And, for the stability of solutions to the least-squares problem, he may refer to van der Sluis (1975). As to problems concerning rank determination, one should not miss the interesting work of Manteuffel (1981). Here A, together with its elements' uncertainties, is contained in an interval matrix A^I, where the author seeks to determine its rank.

Now after making sure that $r(A + \Delta A) = r(A)$ while computing A^+ we are left in the hands of the pseudo-condition number cond (A). This leaves much to be desired, for the latter may be large and the results may still be unsatisfactory.

Among the available methods to improve upon the accuracy of solutions are the iterative refinement techniques. We recall having discussed in detail in sec. 3.6 that if det $(A) \neq 0$ and $x^{(0)}$ is an approximate solution to $Ax = b$, then by defining a sequence of residual vectors $r^{(i)} = Ax^{(i)} - b$, $i = 0, 1, 2, \ldots$ and iteratively solving $A\,\Delta x^{(i)} = r^{(i)}$ and subtracting the result each time from $x^{(i)}$, the latter will ultimately converge to x. To translate this technique onto the least-squares problem would be possible except for one thing: that r will never approach zero. The reason is that Ax and b are never equal, since they are assumed inconsistent from the start ($AA^+b \neq b$). Perhaps only when this overdetermined system is nearly compatible, can this process be stable, since the error in the solution is proportional to $r \cdot \text{cond}^2\,(A)$. But to check that r converges to a minimal value will leave us with an upper bound to check for x (cf. Sect. 3.4), a bound which is again dependent on cond (A).

This problem can be resolved in two ways: either by refining A^+ only and then computing $x = A^+b$, while avoiding the perturbations in b via higher precision multiplication, or by transforming the equations into another set having a larger nonsingular coefficient matrix to which the method of Sect. 3.6 could be adopted. The latter technique was initiated by Björck (1967).

To start with the first option, suppose, for simplicity, that A is of full rank, i.e. $A^+A = I$. Then if $B^{(0)}$ is an approximate Moore-Penrose inverse of A, we perform the following steps of $i = 0, 1, 2$.

a. Form the residual $R^{(i)} = I - B^{(i)}A$, in double precision
b. Calculate $\Delta A^{+(i)} = -R^{(i)}B^{(i)}$
c. $B^{(i+1)} = B^{(i)} - \Delta A^{+(i)}$

But does $B^{(i)}$ approximate A^+ as $i \to \infty$? This we shall see in due course. Let us first proceed with the proof. Similar to the one in Sect. 3.6, the proof for one iteration is as follows:

$$R^{(1)} = I - B^{(1)}A = I - (I + R^{(0)})\,B^{(0)}A = I - (I + R^{(0)})\,(I - R^{(0)}) = (R^{(0)})^2$$

The method converges therefore quadratically with a rapid rate. Let us apply it to the example in Sect. 4.2. We had

$$A = \begin{bmatrix} 0.4087 & 0.1594 \\ 0.4302 & 0.3516 \\ 0.6246 & 0.3384 \end{bmatrix}, \qquad B^{(0)} = \begin{bmatrix} 3.1317 & -2.9327 & 1.5717 \\ -4.4785 & 6.0943 & -1.2669 \end{bmatrix}$$

Hence obtaining in succession using an HP-71 B with ten-digit arithmetic,

$$R^{(0)} = \begin{bmatrix} 3.793 \times 10^{-5} & 8.106 \times 10^{-5} \\ -9.917 \times 10^{-5} & -1.6402 \times 10^{-4} \end{bmatrix}, \quad \text{and}$$

$$B^{(1)} = \begin{bmatrix} 3.131455758 & -2.932317233 & 1.571656920 \\ -4.478076007 & 6.093591249 & -1.266848069 \end{bmatrix}$$

$$R^{(1)} = \begin{bmatrix} -6.6 \times 10^{-9} & 1.2018 \times 10^{-8} \\ -1.2506 \times 10^{-8} & 1.8857 \times 10^{-8} \end{bmatrix}, \quad \text{and}$$

$$B^{(2)} = \begin{bmatrix} 3.131\,455\,783 & -2.932\,317\,276 & 1.571\,656\,923 \\ -4.478\,076\,052 & 6.093\,591\,327 & -1.266\,848\,073 \end{bmatrix}$$

$$R^{(2)} = \begin{bmatrix} -2.9000 \times 10^{-10} & -2.1300 \times 10^{-10} \\ 9.0502 \times 10^{-11} & -1.1417 \times 10^{-10} \end{bmatrix}, \quad \text{and}$$

$$B^{(3)} = \begin{bmatrix} 3.131\,455\,783 & -2.932\,317\,276 & 1.571\,656\,922 \\ -4.478\,076\,052 & 6.093\,591\,327 & -1.266\,848\,073 \end{bmatrix}$$

with $\|R^{(3)}\| \cong 10^{-11}$ and $B^{(4)} = B^{(3)}$ up to ten-digit accuracy; meaning that convergence is reached after three iterations. But does $B^{(3)} \cong A^+$? Of course not; for proposing that the residual R decreases suggests only that $BA \cong I$, i.e. that the matrix B so obtained is only a left-inverse to A. It is true that it will further satisfy $ABA = A$, $(BA)^T = BA$ and $BAB = B$, but not necessarily $(AB)^T = AB$. Therefore Bb will not solve the least-squares problem. We have seen that, in Sect. 4.1, for Bb to be a solution to $A^TAx = A^Tb$, B must satisfy $ABA = A$ and $(AB)^T = AB$. The first condition is satisfied by our algorithm, but the second does not necessarily hold.

To obtain an A^+ of a full rank matrix A, one must refine B so that it satisfies $BA = I$ as well as $(AB)^T = AB$ or the much easier $BA = I$ and $A^T = A^TAB$. The first condition is already dealt with. For the second we shall iterate as follows:

1. Form the residual $R^{(i)} = A^T - A^TAB^{(i)}$ in double precision.
2. Calculate $\Delta A^+ = -B^{(i)}B^{(i)T}R^{(i)}$
3. $B^{(i+1)} = B^{(i)} - \Delta A^+$

To prove that $B^{(i)} \to A^+$ as $i \to \infty$, we have,

$$\begin{aligned} B^{(1)} &= B^{(0)} + B^{(0)}B^{(0)T}R^{(0)} \\ &= A^+A^{+T}(A^T - R^{(0)}) + A^+A^{+T}(A^T - R^{(0)})(A - R^{(0)T})\,A^+A^{+T}R^{(0)} \\ &= A^+ - A^+A^{+T}R^{(0)}A^{+T}R^{(0)} - A^+R^{(0)T}A^+A^{+T}R^{(0)} + A^+A^{+T}R^{(0)}R^{(0)T}A^+A^{+T}R^{(0)} \\ &= A^+ + 0(\|R^{(0)}\|^2, \|R^{(0)}\|^3) \end{aligned}$$

It will be found that,

$$B^{(2)} = A^+ + 0(\|R^{(0)}\|^4, \|R^{(0)}\|^5),$$

meaning that $B^{(i)}$ approaches A^+ by an error of the order of $\|R^{(0)}\|^{2^i}$, thus completing the proof. Obviously, one can carry out both types of iterative schemes simultaneously on the same example; to obtain after one iteration the same $B^{(1)}$ computed above. A second iteration in the other direction was found quite sufficient, as

$$B^{(2)} = \begin{bmatrix} 3.130\,613\,872 & -2.932\,757\,884 & 1.572\,511\,291 \\ -4.476\,577\,940 & 6.094\,375\,353 & -1.268\,368\,353 \end{bmatrix}$$

approximates A^+ by an accuracy of ten significant digits. As to the solution of the least-squares problem, if the perturbations in b are taken into account, the sensitivity of $x = A^+b$ to changes in b is governed by the condition number $\|A^+\| \; \|b\|/\|A^+b\|$ (see exercise 1.21) which is of an order comparative to cond (A). We have therefore succeeded in reducing the error in x — if A is known exactly — from being proportional to cond^2 (A) to an order of cond (A) only.

Björck (1967), on the contrary, does not leave the accuracy of x at the risk of Δb. He manipulated his equations so as to take care of any possible perturbations in either A or b while carrying out the refinement of x. The author proposed to substitute the equations $A^TAx = A^Tb$ by the following equations: $r + Ax = b$, $A^Tr = 0$, which can be combined into $m + n$ equations in the $m + n$ unknowns x and r, namely

$$\begin{bmatrix} I & A \\ A^T & 0 \end{bmatrix} \begin{bmatrix} r \\ \hline x \end{bmatrix} = \begin{bmatrix} b \\ \hline 0 \end{bmatrix}$$

having a nonsingular coefficient matrix for $r(A) = n$, and to which the methods of Sect. 3.6 can be applied.

However, one must not think that subdividing the equations $Ax = b$ into the two sets above leads to a better accuracy. This was asserted by Björck (1967, p. 266) when he deduced that the error in x remains dependent on cond^2 (A). The reader can check the retaining accuracy by comparing the solution of the above two equations with that when solving the normal equations $A^TAx = A^Tb$, for a given choice of b. For the above example, cond $(A) = 9$; the system is therefore stable and the solution comes for instance for $b = (1, 1, 1)^T$ as

$$\hat{x}_1 = 1.770367279\,, \qquad \hat{x}_2 = 3.494290608 \times 10^{-1}$$

with a residual error

$$\hat{r}_1 = 2.207519010 \times 10^{-1}\,, \qquad \hat{r}_2 = 1.155287390 \times 10^{-1}\,,$$
$$\hat{r}_3 = -2.240181963 \times 10^{-1}$$

correct to ten significant digits. To compare this method with the singular value decomposition, have we had to refine either A^+ or x. We supplied the equations of Björck with the initial value of x and r given from $x = B^{(0)}b$ and $r = (I - AB^{(0)})\,b$. One iteration of refinement was found enough to reach the above exact solution, against two iterations for refining A^+ as was seen before. It appears therefore unnecessary to solve the least-squares problem using the singular value decomposition or its variant — the QU decomposition — contrary to wide belief that orthogonalization does not suffer the squaring of cond (A) like when solving $A^TAx = A^Tb$. True that the QU method produces errors proportional to cond (A), but the multiplication of A^+b adds — as we remarked before — another error also proportional to cond (A). This remark came something of a shock — as van der Sluis (1974) truly stated-when Golub and Wilkinson discovered it in 1966. In fact solving the normal equations does produce good results even when cond (A) is large and the data are noisy much over the round-off level, as was shown by Fletcher (1975) when solving them using the stable Cholesky decomposition; a method very well suited to factorize the positive definite matrix of coefficients, since pivoting is also not required and factorization starts from the first entries of $A^T A$. Cholesky's method proved very efficient when handling large-scale sparse systems as outlined in Heath (1984). The latter also surveys and compares different known numerical strategies for handling large normal equations and his article includes an up-to-date bibliography on the subject. The reader interested in a comparison among different refinement techniques may consult Björck (1978).

We should emphasize though that the QU method is the method of choice for the least-squares problem. While $\text{cond}^2(A)$ affects the accuracy of solution of the normal equations, that of the QU method is governed by $\varrho \, \text{cond}^2(A)$, where

$$\varrho = \frac{\|r\|}{\|A\| \, \|x\|}$$

(see Sect. 4.2). Thus for a system in which $\varrho < 1$ the QU should give better results. Furthermore, if $\varrho \, \text{cond}(A) < 1$, or that the first term dominates in the bound for $\|\Delta x\|/\|x\|$ in Sect. 4.2 does the accuracy become affected by $\text{cond}(A)$ only. The orthogonalization method presides therefore over the normal form in terms of accuracy of solution (see Stoer and Bulirsch (1980, p. 208) for comparison).

Summarizing, we quote Noble (1976) in his conclusion. In connection with the Moore-Penrose inverse, the most straightforward procedure for computing A^+ is

$$A^+ = (A^T A)^{-1} A^T (m > n) \quad \text{or} \quad A^T (A A^T)^{-1} (m < n) \, .$$

using double precision if $A^T A$ is badly conditioned. If there is a difficulty concerning rank determination, the singular value decomposition can be used. As for solving the least-squares problem, it is unnecessary and inefficient to first calculate A^+ and then form $A^+ b$. The user concerned with single precision results, is advised to carry an iterative refinement procedure like in Björck (1967). In fact, iterative methods are worth using in large sparse matrix problems. The two equations of Björck, namely

$$r + Ax = b \quad \text{and} \quad A^T r = 0 \, ;$$

furnish the basis of the iterative scheme. They are also used in conjunction with many elimination methods for solving sparse problems, the two equations are combined together to form what is called now the *sparse tableau*. Another advantage associated with the two equations of Björck, is that they can be employed to test the accuracy of solution against a-posteriori measure similar to the one described in Sect. 2.5. When applied here, the latter yields easily the two criteria

$$|A^T \hat{r}| \leqq \varepsilon |A^T| \, |\hat{r}|$$
$$|\hat{r} + A\hat{x} - b| \leqq \varepsilon (|\hat{r}| + |A| \, |\hat{x}| + |b|) \, ,$$

computed in double precision, where $\hat{r}$ and $\hat{x}$ are approximate solutions to r and x. To illustrate these two bounds on the foregoing example, we have

$$|A^T \hat{r}| \cong \begin{bmatrix} 6 \times 10^{-11} \\ 2 \times 10^{-11} \end{bmatrix}, \qquad |A^T|\,|\hat{r}| \cong \begin{bmatrix} 2.798\,435\,308 \times 10^{-1} \\ 1.516\,155\,153 \times 10^{-1} \end{bmatrix}$$

$$|\hat{r} + A\hat{x} - b| \cong \begin{bmatrix} 0 \\ 0 \\ 3 \times 10^{-10} \end{bmatrix}, \qquad |\hat{r}| + |A|\,|\hat{x}| + |b| \cong \begin{bmatrix} 2 \\ 2 \\ 2.448\,036\,392 \end{bmatrix}$$

which satisfy the above two inequalities for a machine precision $\varepsilon = 5 \times 10^{-10}$.

In these last two sections, we have studied perturbation theory of generalized inverses as well as the numerical solution to the least-squares problem. We have also seen that cond (A) affects the desired computational accuracy. We have also thrown some light on some iterative techniques which may be used to improve the results. The latter have attracted workers and still do. As to the problem of scaling for numerical stability, we can mention the bound furnished for the spectral condition number by van der Sluis (1969)

$$\text{cond}_2(A) = \text{cond}_2^{1/2}(A^T A) \leq \sqrt{n} \min_D \text{cond}_2(AD)$$

if all columns of A are scaled down to lenght unity. The reader may also refer to Golub and Van Loan (1983, p. 179) for procedures related to row and column weighing as well as for methods of iterative improvements.

4.4 Case of Interval Coefficients

An experimenter wishing to describe a set of uncertain data (x_i, y_i), $i = 1, \ldots, m$, by a straight-line relation is led to three options. He can monitor x_i to vary among a set of m sample points while making a corresponding observation y_i. Due to the different sources of errors (sampling, human, instrumentation, ...), each observed value y_i will differ from its expected value by an error e_i. Choosing, for instance, the least-squares method to minimize the sum of the squares of these small residuals, he can determine his linear relation. However a different experimenter, knowing that the error e_i is also uncertain, prefers to take several observations y while fixing x_i. He will then notice that most of these observations differ from one another, but naturally lie within a certain range which he may well record. The vector $y\{y_i, i = 1, \ldots, m\}$ will then be an interval vector, A third experimenter, being more sophisticated, notices that although y takes different values at each x_i, it usually centers around a mean one and only differs by a certain variance.

Although the three problems arise from a simple experiment, they are by no means trivial. Theoretically, the first problem (the ordinary least-squares) is the easiest to tackle in the big hierarchy. Practically speaking we are usually faced with problems like data dependence (rank determination see Sect. 4.3) and there still remains a long queue of algorithms when it comes to choose from. The second one possesses one step of difficulty regarding the vast range of solutions obtained using the loose bounds offered by interval mathematics. The third problem, very popular in the circles of econometricians, have kept them busy for almost half a century.

All three problems can be formulated by the general linear model

$$y = X\beta + e$$

where y is a vector of m observations, X is a known $m \times n$ design matrix and e is a vector of m uncorrelated errors. In the ordinary least-squares problem, we seek a vector β under which $\|e\|_2 = \|y - X\beta\|_2$ is minimum, giving simply that $\beta = X^+y$. In the second problem, although we may adopt the least-squares method the solution β will be altered at best to X^+y^I. The third problem can only be resolved through an assumption regarding the nature of the error vector e. This is the central problem of Gauss and we shall return to it in the next section. Indeed many other problems in the above hierarchy are still unresolved; for example, one in which X itself is uncertain, where each x_{ij} has a distribution function with some mean and variance. This problem can only be handled using experimental techniques like the Monte-Carlo method; but generally speaking little if anything is known regarding the distribution of the solution β.

In this section we shall consider the second problem above, but add to it the case in which A is also an interval matrix, i.e.

$$b^I = A^I x + r ,$$

where A^I is a given $m \times n$ interval matrix and b^I is a given $m \times 1$ interval vector. We shall assume for the sake of simplicity, but without loss of generality, that $r(A) = n$, $\forall A \in A^I$, hence that $m \geqq n$. But since we treat least-squares solutions, we take $m > n$ and $r(A) \neq r(A{:}b)$, $\forall A \in A^I$ and $b \in b^I$; i.e. the equations are inconsistent.

When endeavouring to solve the above problem using the least-squares method, we set out to find all possible values of the vector x with the corresponding residual vector r satisfying $b = Ax + r$, where both A and b are fixed and assume all possible combinations of values inside A^I and b^I. In other words, we execute a kind of independent sampling in both A^I and b^I to choose at one time a fixed A and b. We then solve the least-squares problem $b = Ax + r$ by minimizing $\|r\|_2$, giving as a solution

$$x = (A^T A)^{-1} A^T b = A^+ b ,$$

with a corresponding error vector r. Imagining that we can keep repeating the sampling process until we exhaust A^I and b^I, we obtain an infinite number of solutions constituting a region in $\mathbb{R}^n$ which we shall denote X. The latter will therefore contain all least-squares solutions to the linear equations $Ax = b$ where $A \in A^I$ and $b \in b^I$. In other words, solving $A^I x = b^I$ in a least-squares sense is synonymous with finding a set

$$X = \{x : A^T A x = A^T b,\ A \in A^I,\ b \in b^I\}$$

Obviously, related to the set X, there exists a set R containing all possible error vectors r. The problem which we are seeking to solve is to find the vector x^I — one of smallest span — enclosing X. Note that $b^I \neq A^I x^I + r^I$; in fact $b^I \subseteq A^I x^I + r^I$. This incompatibility between both sides is due to two reasons: inconsistency of the equations as well as interval independency (cf. Sect. 2.4).

Unlike the previous case $A^I x = b^I$ in which $\det(A) \neq 0$, $\forall A \in A^I$, treated in chapter two, the above problem is more involved. We remember that X has always been convex in one orthant, and bounded, for instance, by the inequalities $\underline{A}x \leqq \bar{b}$ and $\bar{A}x \geqq \underline{b}$ for $x \geqq 0$. The latter are obtained from the assumption $A^I x \cap b^I \neq \emptyset$ set out by Hansen (see Sect. 2.4). To apply here a similar assumption is straightforward. Unfortunately the two solutions $\underline{A}^T \underline{A} x = \bar{A}^T \bar{b}$ and $\bar{A}^T \bar{A} x = \underline{A}^T \underline{b}$ lying on the boundaries of the set described by $\underline{A}^T \underline{A} x \leqq \bar{A}^T \bar{b}$ and $\bar{A}^T \bar{A} x \geqq \underline{A}^T \underline{b}$ do not belong to X. Another more pronounced complication stems from the fact that X itself is in general a nonconvex set in any one orthant. It cannot be defined therefore by linear inequalities. To show this on a simple example, consider;

$$\begin{bmatrix} 1 & 0 \\ 0 & 1 \\ a & 1 \end{bmatrix} \begin{bmatrix} x_1 \\ x_2 \end{bmatrix} = \begin{bmatrix} 1 \\ 1 \\ 1 \end{bmatrix}, \qquad a \in [0, 1]$$

having as least-squares solution

$$x_1 = \frac{2}{2 + a^2} = [2/3, 1]$$

$$x_2 = \frac{a^2 - a + 2}{2 + a^2} = [2/3, 1]$$

The set X is described by the curve

$$x_1^2 + 2x_2^2 - x_1 - 4x_2 + 2 = 0\,, \qquad 1 \geqq x_1\,, x_2 \geqq \frac{2}{3}\,,$$

which is a nonconvex set. This phenomenon is not encountered in the previous situation ($\det(A) \neq 0$); for when one element in A ranges over a certain interval, X is convex, i.e. a straight-line. This well known fact follows from showing that dx_i/dx_j is independent of a certain a_{kl}. From Sect. 1.6, we have that

$$\frac{dx_i}{dx_j} = \frac{dx_i/da_{kl}}{dx_j/da_{kl}} = \frac{-e_i^T A^{-1}(dA/da_{kl})\, x}{-e_j^T A^{-1}(dA/da_{kl})\, x} = \frac{\text{element}\,(i,k)\text{ in }A^{-1}}{\text{element}\,(j,k)\text{ in }A^{-1}} = \frac{\text{cofactor of } a_{ki}}{\text{cofactor of } a_{kj}}$$

independent of a_{kl}. On the contrary the same quotient becomes generally a function of a_{kl} using the cumbersome formula for dA^+/da_{kl} (see exercise 4.18). The region X is therefore difficult to define. An easier task would be to enclose X in a convex domain defined directly from A^I and b^I. Here again the task is not straightforward. We may suggest the set described for $A, b \geqq 0$, $A \in A^I$, $b \in b^I$ as

$$\underline{A}^T \underline{A} x \leqq \bar{A}^T \bar{b}\,, \qquad \bar{A}^T \bar{A} x \geqq \underline{A}^T \underline{b}$$

But this set leads to a very large overestimation error, because the two points — as we mentioned earlier — $\underline{A}^T\underline{A}x = \overline{A}^T\overline{b}$ and $\overline{A}^T\overline{A}x = \underline{A}^T\underline{b}$ do not lie inside X. While suggesting the set

$$\underline{A}^T\underline{A}x \leqq \underline{A}^T\overline{b}\,, \qquad \overline{A}^T\overline{A}x \geqq \overline{A}^T\underline{b}$$

whose points $\underline{A}^T\underline{A}x = \underline{A}^T\overline{b}$, $\overline{A}^T\overline{A}x = \overline{A}^T\underline{b}$ lie in X, is also wrong by considering the simple counter-example

$$\begin{bmatrix} 1 & 0 \\ 0 & 1 \\ [0,1] & 0 \end{bmatrix} \begin{bmatrix} x_1 \\ x_2 \end{bmatrix} = \begin{bmatrix} 1 \\ 1 \\ 3 \end{bmatrix}$$

And since we cannot dispence with generality by letting A or b take positive or negative values, we must refer back to the two equations set by Björck in Sect. 4.3, i.e. $Ax + r = b$, $A^Tr = 0$. Here they become

$$\begin{bmatrix} I & A^I \\ A^{IT} & 0 \end{bmatrix} \begin{bmatrix} r \\ \hline x \end{bmatrix} = \begin{bmatrix} b^I \\ \hline 0 \end{bmatrix}$$

to which Hansen's inequalities (see Sect. 2.4) can be applied. But the story has not ended, the vector r for a particular choice of A and b is alternating in sign from the very definition of the least-squares problem. But suppose we know some range of r over $x \geqq 0$, then we can pick up a suitable A^I and $(A^I)^T$. For example, take the previous example;

$$\begin{bmatrix} 1 & 0 \\ 0 & 1 \\ [0,1] & 1 \end{bmatrix} \begin{bmatrix} x_1 \\ x_2 \end{bmatrix} = \begin{bmatrix} 1 \\ 1 \\ 1 \end{bmatrix}$$

$r_3 \leqq 0$, hence the set X becomes for $x \geqq 0$

$$\begin{array}{llll} r_1 + x_1 \leqq 1\,, & r_1 + r_3 \leqq 0\,, & r_1 + x_1 \geqq 1\,, & r_1 \geqq 0 \\ r_2 + x_2 \leqq 1\,, & r_2 + r_3 \leqq 0\,, & r_2 + x_2 \geqq 1\,, & r_2 + r_3 \geqq 0 \\ r_3 + x_2 \leqq 1\,, & r_3 + x_1 + x_2 \geqq 1 & & \end{array}$$

giving

$$\begin{array}{ll} 1 \geqq x_1,\ x_2 \geqq 0\,, & x_1 + x_2 \geqq 1 \\ 0 \leqq r_1,\ r_2 \leqq 1\,, & 0 \geqq r_3 \geqq -1 \end{array}$$

satisfied by the exact range $x_1^I = x_2^I = [2/3, 1]$. It can also be defined by the equivalent set (see Sect. 2.4)

$$\begin{array}{lll} |r_1 + x_1 - 1| \leqq 0\,, & |r_2 + x_2 - 1| \leqq 0\,, & |r_3 + (1/2)\,x_1 + x_2 - 1| \leqq (1/2)\,x_1 \\ |r_1 + (1/2)\,r_3| \leqq (1/2)\,|r_3|\,, & |r_2 + r_3| \leqq 0 & \end{array}$$

being usually much more general to apply. The difficulty therefore in applying the bounds of Sect. 2.4 for $x \geqq 0$ stems from the fact that, although $\underline{A}$ is a suitable choice of A^I, it is not for $(A^I)^T$ over the whole range of X. The exact set X is defined correctly for $x \geqq 0$ by the four inequalities

$$r + \underline{A}x \leqq \bar{b}\,, \qquad A_1^T r \leqq 0$$
$$r + \bar{A}x \geqq \underline{b}\,, \qquad A_2^T r \geqq 0$$

where A_1 and A_2 are equal respectively to $\underline{A}$ and $\bar{A}$ except at those elements corresponding to negative values of r_i. For a set of linear inequalities in x alone, one has

$$A_2^T(\underline{A}x - \bar{b}) \leqq 0\,, \qquad A_1^T(\bar{A}x - \underline{b}) \geqq 0\,.$$

These two inequalities are easy to apply but suffer the major drawback of not knowing beforehand A_1 and A_2 to substitute for in the expressions, except in very special cases in which we can guess the signs of r_i. For the above examined example, $r_3 \leqq 0$ and hence we are led to a convex region enclosing X, namely

$$\underline{A}^T\underline{A}x \leqq \underline{A}^T\bar{b} \quad \text{and} \quad \bar{A}^T\bar{A}x \geqq \bar{A}^T\underline{b}\,,$$

defining

$$x^I = ([1/2, 1], [1/2, 1])^T\,.$$

As to the two points $\underline{A}^T\underline{A}x = \underline{A}^T\bar{b}$ and $\bar{A}^T\bar{A}x = \bar{A}^T\underline{b}$ lying in X, they are $x^1 = (1, 1)$, $x^2 = (2/3, 2/3)$, determining x^I exactly.

The above procedure to define x^I is therefore inefficient. A systematic approach could be one of the methods suggested by Hansen and Smith (1967) and discussed in detail in Miller (1972) (see Sect. 2.3). The equations

$$A^{I^T}A^I x = A^{I^T}b^I$$

are rewritten in the form

$$ZA^{I^T}A^I x = ZA^{I^T}b^I$$

where Z is an $n \times n$ matrix equal to the mid-point inverse of $A^{I^T}A^I$. An easier form is also the iterative formula provided in Sect. 3.7.

$$z^{I(k+1)} = \{E^I z^{I(k)} + ZA^{I^T}b^I\} \cap z^{I(k)}\,, \qquad k = 0, 1, 2, \ldots$$

with

$$E^I = I - ZA^{I^T}A^I$$

and

$$z_i^{I(0)} = [-1, 1] \frac{\| Z A^{I^T} b^I \|}{1 - \| E^I \|}, \qquad i = 1, 2, \ldots, n .$$

But because we dispensed with the residual vector r^I, we must expect a loose bound associated with a small rate of convergence. For instance, for the above example,

$$E^I = \begin{bmatrix} [-5/11, 5/11] & [-4/11, 4/11] \\ [-4/11, 4/11] & [-1/11, 1/11] \end{bmatrix};$$

$$Z A^{I^T} b^I = \begin{bmatrix} [4/11, 12/11] \\ [8/11, 10/11] \end{bmatrix}, \qquad z_i^{I(0)} = [-6,6]$$

where the choice of the latter is based on l_∞-norm. Convergence is reached after fifteen iterations with a

$$z^I = ([-2.19, 3.66], [-0.81, 2.46])^T \supset x^I$$

and satisfying

$$z^I = E^I z^I + Z A^{I^T} b^I .$$

An even better method is to iterate simultaneously on both r^I and x^I. E^I and Bb^I (cf. Sect. 3.7) are

$$E^I = \begin{bmatrix} A^c (A^{c^T} A^c)^{-1} (A^{c^T} - A^{I^T}) & A^c (A^{c^T} A^c)^{-1} A^{c^T} A^I - A^I \\ (A^{c^T} A^c)^{-1} (A^{I^T} - A^{c^T}) & I - (A^{c^T} A^c)^{-1} A^{c^T} A^I \end{bmatrix}$$

$$Bb^I = \begin{bmatrix} b^I - A^c (A^{c^T} A^c)^{-1} A^{c^T} b^I \\ (A^{c^T} A^c)^{-1} A^{c^T} b^I \end{bmatrix}$$

which become for the above example

$$E^I = \begin{bmatrix} 0 & 0 & [-4/9, 4/9] & [-1/9, 1/9] & 0 \\ 0 & 0 & [-1/9, 1/9] & [-2/9, 2/9] & 0 \\ 0 & 0 & [-1/9, 1/9] & [-2/9, 2/9] & 0 \\ 0 & 0 & [-4/9, 4/9] & [-1/9, 1/9] & 0 \\ 0 & 0 & [-1/9, 1/9] & [-2/9, 2/9] & 0 \end{bmatrix}, \qquad Bb^I = \begin{bmatrix} 1/9 \\ -2/9 \\ 2/9 \\ 8/9 \\ 7/9 \end{bmatrix}$$

and starting from $z_i^{I(0)} = [-4, 4]$, we obtain after ten iterations a stabilization in the results up to the second decimal, with

$$r_1^I = [-0.29, 0.5], \qquad r_2^I = [-0.57, 0.13], \qquad r_3^I = [-0.13, 0.57]$$
$$z_1^I = [0.49, 1.28], \qquad z_2^I = [0.43, 1.12].$$

One directly notices that the bounds are tighter and very close to being correct notwithstanding the smaller number of iterations executed. This is however achieved at the expense of computational costs. For a discussion of the above method with application to numerical examples, the reader is referred to Spellucci and Krier (1976). The authors calculated also upper and lower bounds for the span of the solution x in relation to those of A and b. The reader should notice that Hansen's method, although adapted here to solve the least-squares problem, can fall into the largest context of minimizing a residual between a true and an expected value of two quantities desired to be compatible in some sense. For instance, instead of minimizing the norm of the residual only, we may further impose a restriction on its width. Such a situation arises when fitting polynomials to a set of points of which a straight-line is a special case. The reader who is interested in methods of fitting interval polynomials may consult the work of Rokne (1978).

The exaggerated bounds obtained by the above method can be further tightened using Hansen 1969's method (see Sect. 2.3). A choice of an $A \in A^I$ and $b \in b^I$ which maximizes x_i is made by watching the signs of $\partial x_i/\partial a_{kl}$ and $\partial x_i/\partial b_k \cdot \bar{x}_i$ is obtained by solving $A^T Ax = A^T b$ in which $a_{kl} = \bar{a}_{kl}$, and $b_k = \bar{b}_k$ for positive signs etc. ... In our problem the signs of

$$\frac{\partial x_i}{\partial a_{kl}} = e_i^T (A^T A)^{-1} \left(\frac{\partial A^T}{\partial a_{kl}} b - \frac{\partial A^T}{\partial a_{kl}} Ax - A^T \frac{\partial A}{\partial a_{kl}} x \right)$$

and

$$\frac{\partial x_i}{\partial b_k} = e_i^T (A^T A)^{-1} A^T \frac{\partial b}{\partial b_k}$$

are noticed and the corresponding choice for a_{kl} and b_k is made. By renewing the calculations for each x_i and each a_{kl} and b_k we determine x^I. The operations are indeed time consuming but pay off in terms of accuracy. Let us illustrate the technique on the example of Jahn (1974)

$$A^I = \begin{bmatrix} [2, 3] & [1, 1] \\ [0, 2] & [0, 1] \\ [0, 1] & [2, 3] \end{bmatrix}, \qquad b^I = \begin{bmatrix} [1, 2] \\ [2, 2] \\ [3, 3] \end{bmatrix}$$

We shall take for A the mid-point A^c and for b, b^c etc. ... Hence we have:

$$\frac{\partial x_1}{\partial a_{11}} = [30 \quad -17] \left\{ \begin{bmatrix} 1.5 \\ 0 \end{bmatrix} - \begin{bmatrix} 1.93 \\ 0 \end{bmatrix} - \begin{bmatrix} 0.78 \\ 0.31 \end{bmatrix} \right\} < 0$$

$$\frac{\partial x_1}{\partial a_{21}} = [30 \quad -17] \left\{ \begin{bmatrix} 2 \\ 0 \end{bmatrix} - \begin{bmatrix} 0.88 \\ 0 \end{bmatrix} - \begin{bmatrix} 0.31 \\ 0.16 \end{bmatrix} \right\} > 0$$

$$\frac{\partial x_1}{\partial a_{12}} = [30 \quad -17] \left\{ \begin{bmatrix} 0 \\ 1.5 \end{bmatrix} - \begin{bmatrix} 0 \\ 1.93 \end{bmatrix} - \begin{bmatrix} 2.89 \\ 1.16 \end{bmatrix} \right\} < 0$$

$$\frac{\partial x_1}{\partial a_{22}} = [30 \quad -17]\left\{\begin{bmatrix}0\\2\end{bmatrix} - \begin{bmatrix}0\\0.88\end{bmatrix} - \begin{bmatrix}1.16\\0.58\end{bmatrix}\right\} < 0$$

$$\frac{\partial x_1}{\partial a_{31}} = [30 \quad -17]\left\{\begin{bmatrix}3\\0\end{bmatrix} - \begin{bmatrix}3.05\\0\end{bmatrix} - \begin{bmatrix}0.16\\0.78\end{bmatrix}\right\} > 0$$

$$\frac{\partial x_1}{\partial a_{32}} = [30 \quad -17]\left\{\begin{bmatrix}0\\3\end{bmatrix} - \begin{bmatrix}0\\3.05\end{bmatrix} - \begin{bmatrix}0.58\\2.89\end{bmatrix}\right\} > 0$$

$$\frac{\partial x_1}{\partial b_1} = [58 \quad 21.5 \quad -27.5]\begin{bmatrix}1\\0\\0\end{bmatrix} > 0$$

Also,

$$\partial x_1/\partial b_2 > 0\,, \qquad \partial x_1/\partial b_3 < 0$$

And it follows that $x_{1\,\max}$ is obtained from solving the least-squares problem

$$\begin{bmatrix}2 & 1\\2 & 0\\1 & 3\end{bmatrix}\begin{bmatrix}x_1\\x_2\end{bmatrix} = \begin{bmatrix}2\\2\\3\end{bmatrix}$$

giving $x_1 = 11/13$, $x_2 = 44/65$. Hence $x_{1\,\max} = 11/13$. This same point coincides with $x_{2\,\min}$ since the signs are all reversed when calculating the partial derivatives of x_2 w.r.t. a_{ij}. As for $x_{1\,\min}$ and $x_{2\,\max}$, we solve

$$\begin{bmatrix}3 & 1\\0 & 1\\0 & 2\end{bmatrix}\begin{bmatrix}x_1\\x_2\end{bmatrix} = \begin{bmatrix}1\\2\\3\end{bmatrix}$$

giving $x_1 = -1/5$, $x_2 = 8/5$. But x_1 with a negative sign suggests the possibility of finding a better point than $-1/5$ to represent $x_{1\,\min}$. This point is not reached possibly because it is far from x^c. Hence using the above point as the new starting point reverses the signs of $\partial x_1/\partial a_{11}$ and $\partial x_1/\partial a_{31}$ only, suggesting that $x_{1\,\min}$ and $x_{2\,\max}$ are obtained from solving

$$\begin{bmatrix}2 & 1\\0 & 1\\1 & 2\end{bmatrix}\begin{bmatrix}x_1\\x_2\end{bmatrix} = \begin{bmatrix}1\\2\\3\end{bmatrix}$$

giving $x_{1\,\min} = -6/14$, $x_{2\,\max} = 25/14$. From which it follows that

$$x^I_{\text{opt}} = ([-6/14,\ 11/13],\ [44/65,\ 25/14])^T$$

The above methods for computing x^I can be applied to calculate $(A^I)^+$. Setting $b = I$ and solving $A^{I^T} A^I X = A^{I^T}$ yields an estimation of $(A^I)^+$. For instance, one can use

$$X^{I(k+1)} = \{(I - ZA^{I^T} A^I)\, X^{I(k)} + ZA^{I^T}\} \cap X^{I(k)}, \qquad k = 0, 1, 2$$

similar to the method used to enclose x^I. A better formula is to iterate on both R^I and X^I where $R + AX = I$, $A^T R = 0$, $A \in A^I$. The latter constitute $m + n$ interval equations of which the inverse of the combined coefficient matrix yields A^{I^+} in its lower left corner. Other methods for evaluating A^{I^+} rely upon an approximation of $(A^{I^T} A^I)^{-1}$ alone using well-known methods (see Sect. 2.3), then performing the multiplication with $(A^I)^T$. This leads usually to overestimated bounds, but could serve as a starting value for $(A^I)^+$ to which further iterative methods are applied.

Naturally we have depicted only a few of the iterative methods for updating an approximate solution; still there remains a great number which we have not mentioned. The reader is referred to an interesting review of this subject by Mönch (1978) who is also dating references from 1965 onwards. His work discusses four methods for obtaining a *monotone enclosure* of A^+ very well suited to $(A^I)^+$. Another relevant method is also Evan's inversion method and is discussed in Alefeld (1984).

One of the methods for finding A^{I^+} is to enclose each of its elements in an optimal interval whose upper and lower bounds are evaluated according to the rule of signs mentioned before. Watching the sign of $\partial x_{ij}/\partial a_{kl}$ determines the value of a_{kl}, i.e. whether we choose $\bar{a}_{kl}$ or $\underline{a}_{kl}$ to ensure a maximum or minimum of x_{ij}. The sign of $\partial x_{ij}/\partial a_{kl}$ is determined from the expression

$$\frac{\partial X}{\partial a_{kl}} = (A^T A)^{-1} \left\{ \frac{\partial A^T}{\partial a_{kl}} - \frac{\partial A^T}{\partial a_{kl}} AX - A^T \frac{\partial A}{\partial a_{kl}} X \right\}$$

for each x_{ij}. For example for

$$A^I = \begin{bmatrix} 1 & 0 \\ 0 & 1 \\ [0, 1] & 1 \end{bmatrix}$$

we have approximately that

$$\frac{\partial X}{\partial a_{31}} = \begin{bmatrix} -32/81 & -28/81 & 28/81 \\ -28/81 & 16/81 & -16/81 \end{bmatrix}$$

Hence a choice of $a_{31} = 0$ maximizes x_{11}, x_{12}, x_{21} and x_{23} while minimizing x_{13} and x_{22}. But since

$$\underline{A}^+ = \begin{bmatrix} 1 & 0 & 0 \\ 0 & 1/2 & 1/2 \end{bmatrix}, \qquad \bar{A}^+ = \begin{bmatrix} 2/3 & -1/3 & 1/3 \\ -1/3 & 2/3 & 1/3 \end{bmatrix}$$

one has

$$A^{I^+} \in \left[\begin{bmatrix} 2/3 & -1/3 & 0 \\ -1/3 & 1/2 & 1/3 \end{bmatrix}, \begin{bmatrix} 1 & 0 & 1/3 \\ 0 & 2/3 & 1/2 \end{bmatrix} \right] = \begin{bmatrix} [2/3, 1] & [-1/3, 0] & [0, 1/3] \\ [-1/3, 0] & [1/2, 2/3] & [1/3, 1/2] \end{bmatrix}$$

and it follows directly, if $\partial X / \partial a_{kl} < 0$, $\forall k, l$ over the whole range of X, that

$$A^{I^+} = [\bar{A}^+, \underline{A}^+] .$$

4.5 Least Squares in Regression

The word *regression* became part of the language of statistics as a result of the early work of Sir Francis Galton in 1886, who plotted the average heights of children against those of their parents. Galton remarked that children of tall parents were not so tall as their parents, while those of short ones were not so short as their parents. He then concluded that human height tends to "regress" back to normal. Nowadays, regression analysis is part of the everyday routine of an experimenter wishing to arrive at some mean or average-like relation between two or more variables drawn from a set of available observations. No wonder that all standard texts in statistics contain at least one chapter on regression and almost all issues of statistical journals include articles circling around this topic.

Work in regression starts from the *bivariate chart*, in which each point; for the case of two variables; represents a pair of measurements (x_i, y_i). Usually the points become scattered around a mean curve simulating a certain profile. In other words, for every x_i, we remark the existence of a large number of values of the variable y_i which accumulate around a certain mean and oscillate above and below it. The study of the behaviour of this mean lies at the core of regression analysis. The latter becomes, therefore, a study of the mean of the values of one variable, at a given value of another variable, and how this mean changes with this other variable. The curve or line representing this mean is called a *regression curve or line*. One must note though, that a curve of regression of y_i on x_i can be different from that of the regression of x_i on y_i.

Linear regression analysis follows three different lines: The case where there is no uncertainty in both x_i and y_i can be treated as before by the least-squares method. When y_i only is uncertain, we encounter one step of difficulty. The case in which both x_i and y_i are uncertain becomes the most difficult to handle. One method of dealing with this problem is explained in Duncan (1974). If x and y are continous variables having a bivariate chart, by drawing horizontal and vertical lines through the diagram to isolate each set of points into a square cell, one can build what is called a *bivariate frequency distribution*. It is easy to imagine such a distribution as looking like a cross-word puzzle in which the cells of the unshaded areas each contain a number representing the number of points inside this cell. It may be found that for a certain x_i lying between two bounds, the number in the corresponding vertical cells will keep increasing to a mean value then decrease again. These numbers represent, therefore, some sort of density function of y over x. One method of obtaining the regression curve is to connect the mean value of the vertical cells, at some mean base value of x_i, with the next point by a straight-line to obtain a corrugated line. The more cells we take with as many measurements as have been taken, we obtain a smooth curve representing our regression curve.

One drawback attributed to this method is the assumption that the observations are error free, and departure from normality is due to the genuine variability in the experimental material itself like in most biological or economic models. In this case the regression curve is an average relationship for the population under study. However this departure could be due as well to some errors in our observation by which; for example; although the expected curve is in fact a straight-line, it is not so due to the lack of accuracy in the measurements.

Gauss was the first scientist to settle this issue in his famous paper *Theoria Motus* around the year 1809. He devised the celebrated least-squares method and applied it to measure the orbits of planets from a set of observations. A brief examination of Gauss's work can be found in Deutsch (1965), and for an account of the historical developments of his work, together with priority dispute with Legendre, the reader may enjoy the interesting articles of Seal (1967) and Stigler (1981). Today the least-squares method is used permanently in geodetic adjustments and astronomical calculations. Golub and Plemmons (1980) reported the success of the method in handling 6,000,000 linear equations with 400,000 unknowns. Their paper discusses a method based on orthogonal decomposition to solve the least-squares problem and quotes many references on its numerical aspects, especially suited to large-scale sparse systems. A recent survey on the history of linear regression quoting extensive bibliography can also be found in Hocking (1983).

Gauss' main contribution to the least-squares problem was based on a set of assumptions by which he deduced axiomatically that the residual error u_i between the observed value y_i and the one of regression y_i^r follows a normal (Gaussian) distribution (see Fig. 1). This assumption, which is possibly not rigorous enough,

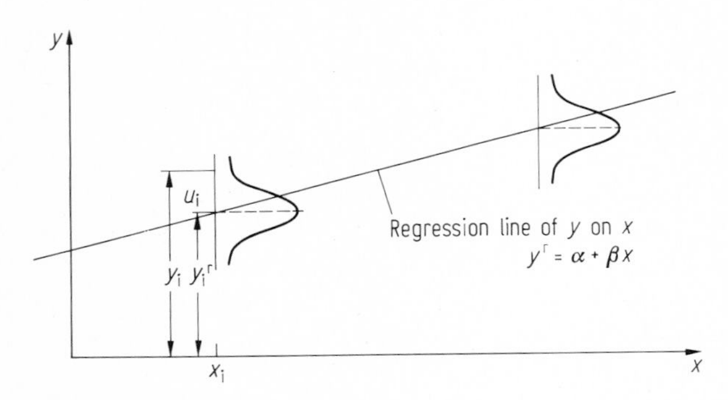

Fig. 4.1

explains the true state of affairs. Since y_i is random, it will indeed center around an expected value and differ by a certain variance. There would be no need to assume a different distribution if we were to take enough measurements with equal probability of error. Econometricians have gained a lot from this assumption and usually assume that the residual u_i has a zero mean and variance σ^2. Moreover the distribution is said to be *homoscedastic* if the standard deviation σ_i of the residuals u_i at each x_i is fixed for all i, that is

$$E(u_i) = 0\,, \qquad \sigma^2(u_i) = \sigma_u^2$$

It helps us, in this respect, to start with the most simple situation in which y_i is free of any uncertainty, in order to draw some general conclusions regarding the classical method of the least-squares. A very good account on the method is to be found in Johnston (1972), and which we shall repeat here. We seek to fit the straight-line $\hat{y} = \hat{a} + \hat{\beta}x$ to a set of points (x_i, y_i). The notations $\hat{\alpha}$ and $\hat{\beta}$ differ from α and β in the sense that the first set is variable according to our possible choice of y_i, while the second set α and β denotes the *regression mean values*. Then by assuming that $y_i = \hat{\alpha} + \hat{\beta}x_i + e_i$, where e_i is a residual error corresponding to the particular choices of y_i, $i = 1, \dots, m$, we obtain, upon minimizing $\sum_{i=1}^{m} e_i^2$, the two equations (cf. Sect. 4.1)

$$\begin{bmatrix} m & \sum x_i \\ \sum x_i & \sum x_i^2 \end{bmatrix} \begin{bmatrix} \hat{\alpha} \\ \hat{\beta} \end{bmatrix} = \begin{bmatrix} \sum y_i \\ \sum x_i y_i \end{bmatrix}$$

to solve for $\hat{\alpha}$ and $\hat{\beta}$. One direct property of the least-squares method is that it allows us to locate the arithmetic mean point $(\Sigma\, x_i/m, \Sigma\, y_i/m)$ on the regression line. This follows easily from the first equation

$$\Sigma\, y_i/m = \hat{\alpha} + \hat{\beta}\, \Sigma\, x_i/m\,.$$

It also follows that $\Sigma\, e_i = 0$, a good result consistent with the very nature of the problem. Now denoting the arithmetic mean of both x_i and y_i, $i = 1, \dots, m$, by $\bar{x}$ and $\bar{y}$, and by defining the new variables

$$\tilde{x}_i = x_i - \bar{x}\,, \qquad \tilde{y}_i = y_i - \bar{y}\,,$$

representing the departure of x_i and y_i from the arithmetic mean, one has simply, from $\Sigma\, \tilde{x}_i = \Sigma\, \tilde{y}_i = 0$, that

$$\hat{\beta} = \frac{\sum_{i=1}^{m} \tilde{x}_i \tilde{y}_i}{\sum_{i=1}^{m} \tilde{x}_i^2}\,, \qquad \hat{\alpha} = \bar{y} - \hat{\beta}\bar{x}$$

Next, let us assume that y_i is uncertain for each x_i and differs from the regression line by a random residual u_i. To exhaust y_i therefore, is to carry repeated sampling for $i = 1, \dots, m$. The value y_i at some x_i will vary therefore from sample to sample as a consequence of different withdrawals from the u distribution in each sample. Hence, applying the above formulae for $\hat{\beta}$ and $\hat{\alpha}$ will generate a series of $\hat{\beta}$ and $\hat{\alpha}$ values pertaining to the distribution of the least-squares estimators; for which we shall calculate the mean and variance. Substituting $y_i = \alpha + \beta x_i + u_i$ in $\hat{\beta}$ and $\hat{\alpha}$ one has:

$$\hat{\beta} = \frac{\sum_{i=1}^{m} \tilde{x}_i \tilde{y}_i}{\sum_{i=1}^{m} \tilde{x}_i^2} = \frac{\sum_{i=1}^{m} \tilde{x}_i y_i}{\sum_{i=1}^{m} \tilde{x}_i^2} = \frac{\sum_{i=1}^{m} \tilde{x}_i(\alpha + \beta x_i + u_i)}{\sum_{i=1}^{m} \tilde{x}_i^2} = \beta + \frac{\sum_{i=1}^{m} \tilde{x}_i u_i}{\sum_{i=1}^{m} \tilde{x}_i^2}$$

and

$$\hat{\alpha} = \bar{y} - \hat{\beta}\bar{x} = \bar{y} - \left(\beta + \frac{\sum_{i=1}^{m} \tilde{x}_i u_i}{\sum_{i=1}^{m} \tilde{x}_i^2} \right) \bar{x} = \alpha + \sum_{i=1}^{m} \left(\frac{1}{m} - \frac{\bar{x}\tilde{x}_i}{\sum_{i=1}^{m} \tilde{x}_i^2} \right) u_i$$

and it follows from $E(u_i) = 0$, that

$$E(\hat{\beta}) = \beta \,, \qquad E(\hat{\alpha}) = \alpha \,;$$

meaning that $\hat{\beta}$ and $\hat{\alpha}$ are *linear unbiased estimates* of β and α. This is indeed a remarkable result concerning least-squares; for increasing the sample size allows the estimated values $\hat{\beta}$ and $\hat{\alpha}$ to approach, in the mean, that of the parent population.

Likewise, the variances of $\hat{\beta}$ and $\hat{\alpha}$ can be obtained in a similar fashion. It is easy to check that

$$\operatorname{var}(\hat{\beta}) = E[(\hat{\beta} - \beta)^2] = \sigma_u^2 \sum_{i=1}^{m} \left(\frac{\tilde{x}_i}{\sum_{i=1}^{m} \tilde{x}_i^2} \right)^2 = \frac{\sigma_u^2}{\sum_{i=1}^{m} \tilde{x}_i^2} \,,$$

by assuming that the residual errors are uncorrelated, i.e. $E(u_i u_j) = 0$, $i \neq j$. Also

$$\operatorname{var}(\hat{\alpha}) = \sigma_u^2 \left(\frac{1}{m} + \frac{\bar{x}^2}{\sum_{i=1}^{m} \tilde{x}_i^2} \right) = \sigma_u^2 \frac{\sum_{i=1}^{m} x_i^2}{m \sum_{i=1}^{m} \tilde{x}_i^2} \,.$$

Here again, we notice that var $(\hat{\alpha})$ decreases with increasing m. In other words, the accuracy in determining α improves the more we take sample points (otherwise the estimator would have values differing systematically from the true value). This is indeed another advantage of the least-squares estimators and is called *consistency* of the estimators. In fact, least-squares estimators are not only consistent unbiased estimators, but also *best linear unbiased*; in the sense that among the class of linear unbiased estimators they have the smallest variances. For supposing that

$$\hat{\beta} = \frac{\sum_{i=1}^{m} (\tilde{x}_i + d_i)\, y_i}{\sum_{i=1}^{m} \tilde{x}_i^2}$$

is also an unbiased estimator, where d_i are some arbitrary constants, it increases the variance of the least-squares estimator by a factor of $\sigma_u^2 \Sigma d_i^2$ which is non-negative.

What remains now is to calculate σ_u^2. A simpler way of determining the latter is via the relation

$$\sigma_u^2 = \frac{E\left(\sum_{i=1}^{m} e_i^2\right)}{m-2};$$

the proof of which results from $\bar{y} = \alpha + \beta\bar{x} + \bar{u}$ obtained by averaging $y_i = \alpha + \beta x_i + u_i$ over the m sample values, as well as $e_i = \tilde{y}_i - \hat{\beta}\tilde{x}_i + (\bar{y} - \hat{\alpha} - \hat{\beta}\bar{x}) = \tilde{y}_i - \hat{\beta}\tilde{x}_i$. It follows, together with subtracting $\bar{y}$ from y_i, that

$$e_i = -(\hat{\beta} - \beta)\,\tilde{x}_i + (u_i - \bar{u});$$

and by taking the expected value of

$$\sum_{i=1}^{m} e_i^2 = (\hat{\beta} - \beta)^2 \sum_{i=1}^{m} \tilde{x}_i^2 + \sum_{i=1}^{m} (u_i - \bar{u})^2 - 2(\hat{\beta} - \beta) \sum_{i=1}^{m} \tilde{x}_i(u_i - \bar{u})$$

results in

$$E\left(\sum_{i=1}^{m} e_i^2\right) = \sigma_u^2 + (m-1)\,\sigma_u^2 - 2\sigma_u^2,$$

coinciding with the above relation. Again, one other remarkable property of the least-squares regression which arose, is that $\Sigma\, e_i^2/(m - 2)$ is an unbiased estimator of σ_u^2.

Now what values can β and α take? Alernatively what is the mean of both populations $\hat{\alpha}$ and $\hat{\beta}$? Since we seldom measure every item in their population, we have to resort to sampling. A good approximation to β and α can thence be obtained from the mean of the sample population. But with how much confidence can we assert such a proposition? Indeed this can be done with the use of confidence intervals. At first, we assume that u_i is normally distributed so are $\hat{\beta}$ and $\hat{\alpha}$, as they form linear functions of u_i. Therefore, at the 95% confidence level, the true mean β and α of the populations $\hat{\beta}$ and $\hat{\alpha}$ is unlikely to lie outside the limits

$$\hat{\beta} \pm 1.96\sigma_{\hat{\beta}}, \qquad \hat{\alpha} \pm 1.96\sigma_{\hat{\alpha}}$$

And yet the sample regression coefficient may be taken to represent an estimate of the true regression coefficient, its accuracy can only be reliable within the confidence levels chosen. Such a study of inferences concerning population by use of samples drawn from it, together with indications of their accuracies, is called *statistical inferences*. The reader interested in a detailed outline on the subject can consult Kendall and Stuart (1961) and Rao (1965).

One major disadvantage attributed to the above confidence limits is that the population standard deviation $\sigma_{\hat{\alpha},\hat{\beta}}$ is unknown and we have to use instead it sample estimate. This will prove satisfactory for large samples only ($m > 30$). But for small samples ($m \leqq 30$), the approximation is poor and small sample theory has to be employed. One very common technique to surmount the difficulty that $\sigma_{\hat{\alpha}, \hat{\beta}}$ is generally unknown and the samples taken small, is the t-distribution. Since we have that

$$\hat{\beta} = N\left(\beta, \frac{\sigma_u}{\sqrt{\sum \tilde{x}_i^2}}\right), \qquad \hat{\alpha} = N\left(\alpha, \sigma_u \frac{\sqrt{\sum x_i^2}}{\sqrt{m \sum \tilde{x}_i^2}}\right)$$

then to apply this test to find confidence limits for β and α, one has that their confidence intervals are respectively given by

$$\hat{\beta} \pm t_{\varepsilon/2, m-2} \frac{\hat{\sigma}_u}{\sqrt{\sum \tilde{x}_i^2}}, \qquad \hat{\alpha} \pm t_{\varepsilon/2, m-2} \hat{\sigma}_u \frac{\sqrt{\sum x_i^2}}{\sqrt{m \sum \tilde{x}_i^2}}$$

where $t_{\varepsilon/2, m-2}$ is the appropriate tabulated value of t with $m - 2$ degrees of freedom corresponding to $100(1 - \varepsilon)\%$ confidence level, and where

$$\hat{\sigma}_u^2 = \frac{\sum_{i=1}^{m} e_i^2}{m - 2}$$

Likewise, the confidence interval for the variate y_k is obtained in the form

$$\hat{\alpha} + \hat{\beta} x_k \pm t_{\varepsilon/2, m-2} \hat{\sigma}_u \sqrt{\frac{1}{m} + \frac{(x_k - \bar{x})^2}{\sum \tilde{x}_i^2}}$$

at a particular x_k. The above expression follows directly from writing

$$\operatorname{var}(\hat{\alpha} + \hat{\beta} x_k) = \operatorname{var}(\hat{\alpha}) + x_k^2 \operatorname{var}(\hat{\beta}) + 2x_k \operatorname{cov}(\hat{\alpha}, \hat{\beta}),$$

with $\operatorname{cov}(\hat{\alpha}, \hat{\beta}) = -\sigma_u^2 \bar{x}/\Sigma\, \tilde{x}_i^2$. Note that since the confidence limits are quadratic functions of x, they form hyperbolic loci around the sample line of regression.

To demonstrate the above results on an example, we depict the one illustrated in Sprent (1969, p. 23). The following observed values of x and y in the table below are obtained by adding random standard normal deviates to values of y corresponding to given x's in accordance with the exact relationship $y = 3 + 2x$.

x	0	1	2	3	4	5	6	7	8	9	10
y	4.28	2.87	6.62	8.9	11.12	15.31	15.47	17.42	17.93	21.22	21.67

Applying the least-squares estimation method, we have $m = 11, \bar{x} = 5, \Sigma\, x_i^2 = 385$, $\Sigma\, \tilde{x}_i^2 = 110$, $\Sigma\, \tilde{x}_i y_i = 215.67$, $\hat{\beta} = 1.9606$, $\hat{\alpha} = 3.0024$, $\hat{\sigma}_u^2 = 1.6518$. Hence the confidence intervals for β, α and y (at some x_k) respectively are:

$$1.9606 \pm 0.277, \quad 3.0024 \pm 1.639$$

and

$$3.0024 + 1.9606x_k \pm 0.876 \sqrt{1 + \frac{(x_k - 5)^2}{10}}$$

where $t = 2.262$ chosen for a 95% confidence level and corresponding to $m - 2$ degrees of freedom. As we said before, the confidence limits are a set of hyperbolic loci for different values of t spread around the line of regression. Other authors suggested, instead of hyperbolas, straight-line segments as confidence limits; as was shown in the classic paper of Graybill and Bowden (1967). A similar suggestion is to be found in Dunn (1968), in which the linear confidence bounds are restricted over a finite range of x.

Generalizations of all above results to the general linear model $y = \beta_1 + \beta_2 x_1 + \dots + \beta_n x_{n-1} + u$ is straightforward. From a set of available measurements $(y_i, x_{1i}, \dots, x_{n-1,i})$, $i = 1, \dots, m$, one can formulate the model in the form

$$y = X\beta + u ,$$

where y is an $m \times 1$ vector of observations, X is a non-stochastic $m \times n$ matrix of known coefficients having a first column with all elements equal to unity, β is $n \times 1$ vector of parameters to be estimated and u is an $m \times 1$ vector of random errors, with

$$E(u) = 0 , \qquad V(u) = E(uu^T) = \sigma^2 I ;$$

i.e. that the errors are uncorrelated ($E(u_i u_j) = 0$, $i \neq j$) but having a homoscedastic distribution. We shall likewise assume that y is a random vector which we shall exhaust by repeated sampling. Also by letting $y = X\hat{\beta} + e$ be one such sample, we have from minimizing $e^T e$ (see Sect. 4.1) that

$$\hat{\beta} = (X^T X)^{-1} X^T y .$$

To show that $\hat{\beta}$ is a best linear unbiased estimator of β, we have from

$$\hat{\beta} = (X^T X)^{-1} X^T (X\beta + u) = \beta + (X^T X)^{-1} X^T u ,$$

that $E(\hat{\beta}) = \beta$. As for the dispersion matrix of $\hat{\beta}$, it is given by

$$\begin{aligned} V(\hat{\beta}) &= E[(\hat{\beta} - \beta)(\hat{\beta} - \beta)^T] \\ &= E[(X^T X)^{-1} X^T uu^T X (X^T X)^{-1}] \\ &= \sigma^2 (X^T X)^{-1} \end{aligned}$$

Next to show that $\hat{\beta}$ is a minimum variance estimator of β, we let $\hat{v} = Ty$ be another linear unbiased estimator of the parameters $C\beta$, i.e. that $E(\hat{v}) = E(Ty) = C\beta$. This implies that $TX = C$, since $E(T(X\beta + u)) = C\beta$. Then calculating $V(\hat{v})$, we have

$$\begin{aligned} V(\hat{v}) &= E[(\hat{v} - C\beta)(\hat{v} - C\beta)^T] = E(Tuu^T T^T) = \sigma^2 TT^T \\ &= \sigma^2[(C(X^T X)^{-1} X^T)(C(X^T X)^{-1} X^T)^T + (T - C(X^T X)^{-1} X^T) \\ &\quad \times (T - C(X^T X)^{-1} X^T)^T] \end{aligned}$$

where the last identity follows by carrying mere multiplications of brackets. And since each of the last two terms are of the form BB^T (positive definite) they have non-negative diagonal elements. But only the second term is a function of T, and it follows that var $(\hat{v})$ is minimum whenever TT^T has minimum diagonal elements, or that $T = C(X^TX)^{-1} X^T$, so that the minimum unbiased estimator of $C\beta$ is $\hat{v} = C(X^TX)^{-1} X^Ty = C\hat{\beta}$ and for $C = I$ the least-squares estimator $\hat{\beta}$ is recovered.

Although the above proof relies upon the sufficient condition that $E(uu^T) = \sigma^2 I$ for $\hat{\beta}$ to be best linear unbiased estimator of β, it is by no means necessary. A less restrictive condition, which was found to be both necessary and sufficient, was discovered by McElroy (1967), by which $\hat{\beta}$ is the best linear unbiased estimator of β if and only if the errors have equal variances and equal non-negative correlation coefficients.

Finally, to estimate σ^2 from the sum of squared residuals e^Te, we have

$$e = y - X\hat{\beta} = (X\beta + u) - X(X^TX)^{-1} X^T(X\beta + u) = (I - X(X^TX)^{-1} X^T)\, u\,.$$

Hence

$$e^Te = u^T(I - X(X^TX)^{-1} X^T)\, u = \operatorname{tr}\,[(I - X(X^TX)^{-1} X^T)\, uu^T]\,.$$

And it follows that

$$E(e^Te) = \sigma^2 \operatorname{tr}\,(I - X(X^TX)^{-1}X^T) = \sigma^2(m - n)\,,$$

generalizing the foregoing result for $n = 2$. In other words, an unbiased estimator of σ^2 is equal to the sum of squared residuals divided by the number of observations minus the number of parameters estimated.

In analysing the above general model, it was assumed that the errors are independent random variables with mean zero and constant variance σ^2. Instead, if the errors are correlated, i.e. $E(uu^T) = \sigma^2\Omega$, the dispersion matrix $V(\hat{\beta})$ becomes equal to $\sigma^2(X^TX)^{-1}X^T\Omega X(X^TX)^{-1}$. But because the matrix Ω is generally unknown, the estimation problem becomes very difficult as Kendall and Stuart (1961, p. 87) reported. A multitude of methods exist to tackle this problem which follow two main approaches. One is estimation by quadratic function of random variables using analysis of variance techniques and without making any assumption about the distribution of the variables. Another, is estimation by the method of maximum likelihood assuming normality of the variables. Apparently, as was stated in Rao (1972), the first method lacks a great deal of theoretical basis and depends much on intuition, while the second is computationally very complicated. The review paper of Searle (1971) surveys many of the classical methods and contains an extensive bibliography. Other recent references are Ljung and Box (1980) and Dempster et al. (1981).

Another remark concerning the above model is the assumption that the matrix X^TX is invertible. The estimation problem when $r(X) < n$ makes use of X^+. For an application of the generalized inverse in statistical inferences, the reader is referred to Rao and Mitra (1971, 1972), Albert (1972, 1976) and Chipman (1976).

At last the concept of confidence intervals, like in the foregoing two parameters' case, can be extended readily to the n-general case of the model $y = X\beta + u$. From the dispersion matrix $V(\hat{\beta}) = \sigma^2(X^T X)^{-1}$, one can obtain var $(\hat{\beta}_i)$, then use the confidence limits set before. However, in many applications, multidimensional confidence regions of several parameters are often simultaneously sought. For example a confidence region regarding $\beta_1^2 + \beta_2^2$ etc. ... as Kendall and Stuart (1961, p. 127) envisioned a topic of considerable difficulty. The reason for seeking a simultaneous confidence region, is that while $\hat{\beta}_1$, obtained for one sample may not be dispersed from β_1, the associated $\hat{\beta}_2$ may well be from β_2. The reader, interested in knowing about multidimensional confidence regions of a general linear model with autocorrelated disturbances, is referred to a method outlined in Deutsch (1965, p. 163).

Finally, we come to the most perplex situation, that in which both variables x_i^r and y_i^r, are subjected to errors. The latter constitute the true variables satisfying the regression relationship and are unobservable. Instead, we observe another pair of observations (x_i, y_i) related to the true variables by some errors, namely,

$$x_i = x_i^r + v_i\,, \qquad y_i = y_i^r + u_i\,.$$

We shall assume like before that $E(u_i) = E(v_i) = 0$ and $\sigma^2(v_i) = \sigma_v^2$, $\sigma^2(u_i) = \sigma_u^2$. Moreover, the errors are uncorrelated, that is cov $(u_i, u_j) =$ cov $(v_i, v_j) = 0$, $i \neq j$ and cov $(v_i, u_i) = 0$. Obviously if $v_i = 0$ we end up with the foregoing simple model $y_i^r = \alpha + \beta x_i$, $y_i^r = y_i - u_i$ or that $y_i = \alpha + \beta x_i + u_i$ to which the least-squares method yields unbiased estimates for α and β. The difficulty inherent in the above model, in which the measurement pair (y_i, x_i) of the dependent as well as independent variables are both uncertain, lies in the fact that x_i is correlated with $(u_i - \beta v_i)$, a situation not to be encountered when $v_i = 0$. For suggesting that the model is given by $x_i = x_i^r + v_i$, $y_i = y_i^r + u_i$ and $y_i^r = \alpha + \beta x_i^r$, gives $y_i = \alpha + \beta x_i + (u - \beta v_i)$ in which the error term $(u_i - \beta v_i)$ is correlated with x_i with a covariance of:

$$\text{cov}\,(x_i, u_i - \beta v_i) = E[x_i(u_i - \beta v_i)] = E[(x_i^r + v_i)(u_i - \beta v_i)] = -\beta\sigma_v^2\,,$$

being only zero of $\sigma_v = 0$. This had led to much confusion among workers for quite a time and raised up the question whether this is at all a problem of linear regression. In all cases, applying the least-squares method to minimize the error term yields a regression line of a smaller slope than the true one as well as biased estimates for β and α. This follows from the relationship:

$$E(\hat{\beta}) = \frac{\text{cov}\,(x^r, y^r)}{\sigma_x^2 r + \sigma_v^2}$$

against,

$$\beta = \frac{\text{cov}\,(x^r, y^r)}{\sigma_x^2 r}$$

for the true regression coefficient of $y^r = \alpha + \beta x^r$. Hence $\hat{\beta}$ is an unbiased estimator of β only if $\sigma_v = 0$. Lindley (1947) has shown that this is not the whole story. For even if the true regression of y^r on x^r is linear, it does not follow for y on x. The latter is only true if the c.g.f. of x^r is a multiple of that of v (see also Kendall and Stuart (1961, ch. 29)). Linearity in regression is thus maintained if we add a further assumption of normality of v_i, a rather very early assumption in regression analysis.

Among the early attempts to reduce the above model to one of regression are those of Wald (1940) and Berkson (1950). The first author proposed a kind of grouping of the observations being independent of the errors. The second assumed x_i to be fixed during the observations y_i, and it is x_i^r which becomes random around x_i with error v_i, i.e. x_i is independent of v_i. Both assumptions lack reality and the problem remains difficult to solve. For an old survey of the available methods until 1960 is the classical paper of Mandasky (1959), see also Moran (1971). But perhaps, the problem is not really one of linear regression as Kendall and Stuart (1961, ch. 29) intuitively asserted. For sure, there is no basis in minimizing errors in the vertical directions (y-direction) as the classical least-squares suggests. Since x_i is also random, one may equivalently minimize in the x-direction, if we are to estimate β correctly. To use least-squares, therefore, one must take into account both vertical and horizontal errors, as Lindely (1947) suggested, i.e. minimizing $\Sigma\, w_i(\hat{\beta})\,(y_i - \hat{\alpha} - \hat{\beta}x_i)^2$, where the weights $w_i(\hat{\beta})$ are proportional to the reciprocals of the variances of $y_i - \hat{\alpha} - \hat{\beta}x_i$. Lindely pointed out that this is exactly equivalent to minimizing the distance between (x_i, y_i) and (x_i^r, y_i^r) for all i. The estimates obtained will also coincide with those deduced from using the maximum likelihood method as was first shown in Mandasky (1959) based only on some assumptions regarding one of the unknowns σ_u^2, σ_v^2 or σ_u^2/σ_v^2.

But if none of the unknowns is known beforehand, the problem has obviously no satisfactory answer. We may not even accept a regression line to fit all available observations. The reason why there would be no satisfactory linear relation among the unobservables as a regression line was furnished geniously by Kendall and Stuart (1961, p. 385). Their argument runs as follows: each pair (x_i, y_i) emanates from an unknown true point (x_i^r, y_i^r). So if we knew σ_u^2 and σ_v^2 we could draw elliptic confidence regions for x_i^r, y_i^r appropriate to any desired probability level and centered at (x_i, y_i). The problem of estimating β and α is therefore a line which intersects as many of these ellipses as possible. And not knowing σ_u^2 and σ_v^2 we neither know the area of these ellipses nor their eccentricity, hence admitting of no one satisfactory regression line. The authors (see p. 413) pushed their dissatisfaction a little further. It looks as though, even if we accept a regression line between x and y, it does not necessarily entail that the true relationship between x^r and y^r is linear. In fact the effects of errors in x^r and y^r is not only responsible of diminishing the slope of the regression lines as we previously showed, but also in impairing the linearity assumption.

However, not many workers are willing to give up easily the method of least-squares, in virtue of its greater simplicity over other methods (see for instance Ketellapper (1983)). Also a simple but most interesting study on the effect of data uncertainties on the least-squares estimators was carried out by Hodges and Moore (1972). The authors favored the least-squares method for its great simplicity and showed that the errors can be easily accounted for. By incorporating the effect of the errors in the independent variables x, while performing a least-squares sampling yields estimators very close to the true ones. This can be done, as the authors explained by considering the model $y^r = X^r\beta$, in which both variables y^r and X^r are subjected to errors and are measured by another uncertain pair y and X related to the true variables by the relationships $y = y^r + u$, $X = X^r + V$. And by minimizing the residual $e = y - X\hat{\beta}$ over the m sampled values (y_i, $x_i's$), we have:

$$\hat{\beta} = (X^TX)^{-1}X^Ty = [(X^r + V)^T (X^r + V)]^{-1} (X^r + V)^T y$$
$$\cong [X^{r^T}X^r + X^{r^T}V + V^TX^r]^{-1} (X^r + V)^T y$$

by neglecting the second order term V^TV. While expanding the inverse of the above matrix (cf. Sect. 1.4) as

$$(X^{r^T}X^r)^{-1} - (X^{r^T}X^r)^{-1} (X^{r^T}V + V^TX^r) (X^{r^T}X^r)^{-1}$$

approximately, gives

$$\hat{\beta} = \hat{\beta}_0 + (X^{r^T}X^r)^{-1} V^Te - (X^{r^T}X^r)^{-1} X^{r^T}V\hat{\beta}_0$$

where $\hat{\beta}_0 = (X^{r^T}X^r)^{-1}X^{r^T}y$ is the estimator had no uncertainty in X^r existed, $e = y - X^r\hat{\beta}_0$ is the residual error vector, again if X^r is accurately known. In the above formula, terms in V^2 are neglected. The sensitivity of $\hat{\beta}_k$ to X^r_{ij} can now be obtained easily in the form:

$$\frac{\partial\hat{\beta}_k}{\partial x^r_{ij}} = (X^{r^T}X^r)^{-1}_{kj} e_i - \left(\sum_{l=1}^{n} (X^{r^T}X^r)^{-1}_{kl} X^r_{il}\right) \hat{\beta}_{0j}$$

showing that the slope of the regression line is not constant as it is originally anticipated, in contrast with the constant one for constant X. But if second order terms in V^2 are considered while calculating $\hat{\beta}$, i.e. the quadratic term in VV^T is taken into account, this will yield a coefficient curve for the estimators very close to the actual one. The authors went on showing easily, by taking $y = X^r\beta + u$, where β stands for the true regression estimator; while assuming that $E(V^TV)/m = \text{diag}\,(\sigma_1^2, \ldots, \sigma_n^2) = S$ and $\sigma_{u_i}^2 = \sigma_u^2$, that

$$E(\hat{\beta}) = \beta - (m - n - 1) (X^{r^T}X^r)^{-1} S\beta$$

giving a biased estimate of β by a term proportional to σ_i^2. Also the variance-covariance matrix of $\hat{\beta}$ is

$$E[(\hat{\beta} - \beta) (\hat{\beta} - \beta)^T] = (X^{r^T}X^r)^{-1} \left(\sigma_u^2 + \sum_{i=1}^{n} \beta_i^2\sigma_i^2\right)$$

reducing to the foregoing least-squares variance $\sigma_u^2(X^TX)^{-1}$ when X^r is error free. Although, the authors concluded that their method has not presented a rigorous examination of the effect of data uncertainties on least-squares regression, in our view the latter could not find someone forwarding a better defense.

Exercises 4

1. If $A^+ABB^*A^* = BB^*A^*$ and $BB^+A^*AB = A^*AB$, show that $(AB)^+ = B^+A^+$.

2. Show that $x = A^+b$ is the unique solution to $Ax = b$ having minimal norm. Hint: Apply the Pythagorean theorem $\|x\|_2^2 = \|A^+b\|_2^2 + \|(I - A^+A)\,c\|_2^2$ are A^+b and $(I - A^+A)\,c$ orthogonal?

3. A norm in $\mathbb{C}^{m \times n}$ is a function $\|\cdot\|: \mathbb{C}^{m\times n} \to \mathbb{R}$ satisfying $\|A\| > 0$ if $A \neq 0$, $\|\alpha A\| = |\alpha|\,\|A\|$ (α scalar) and $\|A + B\| \leqq \|A\| + \|B\|$. Show that if U and V are unitary, that $\|U^*AV\|_2 = \|A\|_2$. Hint: $\|A\|_2 = \sup\limits_{\|x\|_2 = 1} \|Ax\|_2$

4. Show that $\|A\|_2 = \sigma_1(A)$, $\|A^+\|_2 = 1/\sigma_r(A)$, $\sigma_1(A) > \sigma_2(A) > \dots > \sigma_r(A)$, hence $\text{cond}_2\,(A) = \sigma_1/\sigma_r$.

5. If $\sigma_1(A) \geqq \sigma_2(A) \geqq \dots \geqq \sigma_r(A) > 0$ are the singular values of A with rank r, show that

   $$\sigma_i(A) - \sigma_1(E) \leqq \sigma_i(A + E) \leqq \sigma_i(A) + \sigma_i(E)$$

   $$\sigma_i(AC) \leqq \sigma_i(A)\,\sigma_1(C)\,, \quad \sigma_1(A)\,\sigma_i(C)$$

   Hint: $\sigma_i(A) = \sup\limits_{\dim(\mathscr{L}) = i,} \ \inf\limits_{\substack{x \in \mathscr{L} \\ \|x\|_2 = 1}} \|Ax\|_2\,, \qquad i = 1, \dots, r$

6. Show that $\|A\|_F^2 = \Sigma\,\sigma_i^2(A)$. Hint: $\|A\|_F^2 = \text{tr}\,(A^*A)$, use singular value decomposition for A. Obtain also an expression for $\text{cond}_F\,(A)$.

7. If A and B are $m \times n$ matrices with $B - A = E$, denote their respective singular values by α_i and β_i, $i = 1, \dots, k = \min\,(m, n)$. Show that

   $$\sum_{i=1}^{k} (\beta_i - \alpha_i)^2 \leqq \|E\|_F^2\,.$$

8. Show that, if $\det\,(A) = 0$ and $\det\,(A + \varepsilon B) \neq 0$,

   $$(A + \varepsilon B)^{-1} = \frac{C_{n-1} + \varepsilon C_{n-2} + \dots + \varepsilon^{n-2} C_1 + \varepsilon^{n-1} C_0}{a_{n-1}\varepsilon + a_{n-2}\varepsilon^2 + \dots + a_1 \varepsilon^{n-1} + a_0 \varepsilon^n}$$

   where $C_{n-1} = A^a$, $a_{n-1} = \text{tr}\,(A^a B)$, ... Hint: $(\lambda I + A + \varepsilon B)^{-1} = [I + \varepsilon(\lambda I + A)^{-1} B]^{-1}\,(\lambda I + A)^{-1}$; cf. ex. 1.10.

9. If A is an $m \times n$ matrix of rank r and $r(A + \Delta A) > r(A)$, show that $\|(A + \Delta A)^+\|_2 \geqq 1/\|\Delta A\|_2$. Hint: use ex. 5, by letting $i > r$, then $\sigma_i(A + \Delta A) \leqq \|E\|_2$.

10. If A is an $m \times n$ matrix of rank r, then if $\|A^+\|_2\,\|\Delta A\|_2 < 1$, show that $r(A + \Delta A) \geqq r(A)$. Hint: $\sigma_j(A + E) \geqq \dfrac{1}{\|A^+\|_2} - \|E\|_2 \geqq 0$, for some j from ex. 5.

11. If A and B are not in the acute case, show that $\|B^+ - A^+\| \geqq 1/\|B - A\|_2$. Hint: Choose $y \in N(A^*)$, i.e. $A^*y = 0$, then $A^+y = 0$ and $(B^+ - A^+)\,y = B^+y$, $(B - A)^*y = B^*y$, $\|B^+ - A^+\|_2 \geqq \|B^+y\|_2$, $\|B - A\|_2 \geqq \|B^*y\|_2$. Show that $\|B^+ - A^+\|_2\,\|B - A\|_2 \geqq y^*BB^+y = 1$ by further imposing that $y \in R(B)$. Hence obtain a bound for $\|(A + \Delta A)^+ - A^+\|_2^{-1}$ in terms of $\|\Delta A\|_2$.

12. If A is an $m \times n$ matrix of rank k, show that there exists an E with $\|E\|_2 = 1/\|A^+\|_2$ such that $A + E$ has rank $k - 1$. This result is due to Mirsky. Hint: Use the singular value decomposition for $A = U \operatorname{diag}(\sigma_1, \ldots, \sigma_k, 0, \ldots, 0) V^*$. Choose $E = U \operatorname{diag}(0, \ldots, 0, -\sigma_k, 0, \ldots, 0) V^*$. Show further that a perturbation E such that $\|E\|_2 \leqq 1/\|A^+\|_2$ cannot decrease $r(A + E)$.

13. If $\tilde{v}^i = v^i + \varepsilon v^i_{(1)} + \varepsilon^2 v^i_{(2)} + \ldots$ is an eigenvector of $(A + \varepsilon A_1)^* (A + \varepsilon A_1)$ corresponding to the eigenvalue $\tilde{\lambda}_i = \varepsilon^4 \lambda_i^{(4)} + \varepsilon^5 \lambda_i^{(5)} + \ldots$, show that $Av^i_{(1)} + A_1 v^i = 0$. Hint: Premultiply by $((A + \varepsilon A_1)^+)^*$ the eigenvalue problem $(A + \varepsilon A_1)^* (A + \varepsilon A_1) \tilde{v}^i = \tilde{\lambda}_i \tilde{v}^i$. And if A is normal, does $(A + \varepsilon A_1)$ has the eigenvalue $\lambda_i \cong \varepsilon^2 \sqrt{\lambda_i^{(4)}}$? Search for a corresponding eigenvector.

14. If $AA^+ \Delta A = \Delta A$, $A^+A \Delta A^* = \Delta A^*$ and $\|A^+\| \|\Delta A\| < 1$, show that

$$\frac{\|(A + \Delta A)^+ - A^+\|}{\|A^+\|} \leqq \operatorname{cond}(A) \frac{\|\Delta A\|/\|A\|}{1 - (\operatorname{cond}(A) \|\Delta A\|/\|A\|)}$$

with $\operatorname{cond}(A) = \|A\| \|A^+\|$.

15. If $AA^+B = B$ and $\|A^+B\| < 1$, show that $r(A + B) = r(A)$. Hint: $r(A + B) = r[A(I + A^+B)]$. Show using the example $A + \varepsilon A_1 = \begin{bmatrix} 1 & 0 & 0 \\ 0 & 1 & 0 \\ 0 & \varepsilon & 0 \end{bmatrix}$, that the opposite is not true, i.e. that $AA^+A_1 \neq A_1$

16. If A and A_1 commute and are both normal, show that for $r(A + \varepsilon A_1) = r(A)$, $(A + \varepsilon A_1)^+ = A^+ \sum_{k=0}^{\infty} (-)^k (A^+A_1)^k \varepsilon^k$ in the neighbourhood of $\varepsilon = 0$.

17. Show that $B^+ - A^+ = -B^+ \Delta A A^+ + B^+(I - AA^+) - (I - B^+B) A^+$, where $B = A + \Delta A$. Hint: $B^+ - A^+ = (B^+B + I - B^+B)(B^+ - A^+)(AA^+ + I - AA^+)$.

18. Let $A(t)$ be an $m \times n$ matrix with $m \geqq n$ whose elements are differentiable functions of t. Then if $r(A(0)) = n$, show that

$$\frac{dA^+}{dt} = -A^+ \frac{dA}{dt} A^+ + A^+A^{+*} \left(\frac{dA}{dt}\right)^* (I - AA^+)$$

around $t = 0$. Hint: $A(\Delta t) = A(0) + \Delta t \dfrac{dA(0)}{dt} + 0(\Delta t^2)$, hence, apply the Wedin decomposition theorem for $A^+(\Delta t) - A^+(0)$ whose last term vanishes as $A(0)$ is of full rank, then divide by Δt and take limit as $\Delta t \to 0$.

19. If A is an $m \times k$ matrix having rank k, and define $\operatorname{cond}_2(A) = \|A\|_2 \|A^+\|_2$, show that $\operatorname{cond}_2(A^*A) = \operatorname{cond}_2^2(A)$. Hint: A^*A has singular values $\sigma^2(A)$.

20. The approximate solution x of $Ax = b$, with $r(A) \neq r(A \mathrel{\vdots} b)$, is obtained through least-squares fitting, i.e. by minimizing $\|Ax - b\|_E$. Hint: Differentiate with respect to x the quadratic form $(Ax - b)^T (Ax - b)$ to obtain $x = (A^TA)^{-1} A^Tb$. If A^TA is nonsingular, show that under small perturbations ΔA and Δb, the

variation in $x(\Delta x)$ is given by $\Delta x = -(A^T A)^{-1} A^T \Delta A x + (A^T A)^{-1} \Delta A^T (b - Ax) + (A^T A)^{-1} A^T \Delta b$. Hence obtain:

$$\|\Delta x\| \leq \|(A^T A)^{-1} A^T\| \, \|A\| \frac{\|\Delta A\|}{\|A\|} \|x\| + \|(A^T A)^{-1}\| \, \|A^T\|^2 \frac{\|\Delta A^T\| \, \|b - Ax\|}{\|A^T\| \, \|A^T\|}$$

$$+ \|(A^T A)^{-1} A^T\| \, \|A\| \frac{\|b\| \, \|\Delta b\|}{\|A\| \, \|b\|}$$

And by choosing a unitary matrix P such that $PA = \begin{bmatrix} \tilde{A} \\ \hline 0 \end{bmatrix}$; $A = P^T \begin{bmatrix} \tilde{A} \\ \hline 0 \end{bmatrix}$ obtaining $A^T A = \tilde{A}^T \tilde{A}, (A^T A)^{-1} = \tilde{A}^{-1} (\tilde{A}^T)^{-1}, (A^T A)^{-1} A^T = (\tilde{A}^{-1} \vdots 0) \, P$, show that the above bound can be simplified to yield

$$\frac{\|\Delta x\|}{\|x\|} \leq \text{cond}\,(\tilde{A}) \frac{\|\Delta A\|}{\|A\|} + \text{cond}^2\,(\tilde{A}) \frac{\|b - Ax\| \, \|\Delta A\|}{\|A\| \, \|x\| \, \|A\|} + \text{cond}\,(\tilde{A}) \frac{\|b\| \, \|\Delta b\|}{\|A\| \, \|x\| \, \|b\|}$$

by defining a suitable norm.

21. If cond $(A^T A) = \text{cond}^2\,(\tilde{A})$ as in exercise 20, show that for the two expressions $(A^T A + G)(x + \Delta x)$ and $A^T b + \Delta b$ to be equal, then

$$\frac{\|\Delta x\|}{\|x\|} \leq \text{cond}^2\,(\tilde{A}) \left(\frac{\|G\|}{\|A^T A\|} + \frac{\|\Delta b\|}{\|A^T b\|} \right)$$

Hint: make use of the results of sec. 1.4.

22. Let $A \in \mathbb{R}^{m \times n}$ is of full rank and having that $D = \text{diag}\,(1, \ldots, \sqrt{1 + \delta}, \ldots, 1)$, with $\delta > -1$, show that the residual $r(\delta) = b - Ax(\delta)$ for the least-squares solution of $\|D(Ax - b)\|_2$ is given by

$$r_k(\delta) = \frac{r_k(0)}{1 + \delta e_k^T A (A^T A)^{-1} A^T e_k}$$

Hence show how δ can be chosen for an optimum row-scaling of A.

23. If $A \in \mathbb{R}^{m \times n}$ has rank n and

$$M = \begin{bmatrix} \alpha I_m & A \\ A^T & 0 \end{bmatrix}, \qquad \alpha \geqq 0$$

show that $\sigma_{m+n}(M) = \min \left(\alpha, -\frac{\alpha}{2} + \sqrt{\sigma_n^2(A) + \left(\frac{\alpha}{2}\right)^2} \right)$. What value of α minimizes $\text{cond}_2\,(M)$? (Golub and van Loan (1983, p. 183))

24. Obtain an A^{I^+} for $A^I = \begin{bmatrix} [2, 3] & -1 \\ 1 & [0, 1] \\ 1 & -2 \end{bmatrix}$

Chapter 5

Sensitivity in Linear Programming

5.1 Introduction

Whilst the theory of linear equations is concerned with solving the equations $Ax = b$ and the methods involved therein, linear programming is used to study the set of linear inequalities $Ax \leqq b$. These latter inequalities define the set $X = \{x: Ax \leqq b,\ x \geqq 0\}$ in $\mathbb{R}^n$; $n = \dim x$; X being a convex set with a polyhedral shape. Linear programming's major concern becomes then the selection among the vertices of X of the one that would either maximize or minimize the linear function

$$\sum_{i=1}^{n} c_i x_i$$

Mathematically, the problem of linear programming is formulated as follows

$$\max_{x_1, \ldots, x_n} \; c_1 x_1 + c_2 x_2 + \ldots + c_n x_n$$

subject to the restrictions

$$a_{11} x_1 + \ldots + a_{1n} x_n \leqq b_1$$

$$\cdots\cdots\cdots\cdots\cdots\cdots$$

$$a_{m1} x_1 + \ldots + a_{mn} x_n \leqq b_m$$

with

$$x_1, x_2, \ldots, x_n \geqq 0$$

The coefficients $c_1, \ldots, c_n$ are called the *cost coefficients*. The elements a_{ij} are sometimes called the *input-output coefficients*; they form the so-called *technological matrix*. The different values of b define a bound on the available resources that set a limit on production. This nomenclature originated from the practical applications of linear programming. The reader is referred to Gass (1969) for an account on these applications.

Historically, the linear programming problem has been solved by a systematic procedure called the *simplex method*, devised by Dantzig in 1947. This method we will therefore outline briefly. First, the inequalities $Ax \leqq b$ are transformed into equations of the form $Ax + y = b$. This is done by adding a set of new variables $y_j \geqq 0$ with j ranging from 1 to m. The variables y_j are called *slacks*, and are used to balance both sides of the inequalities into equations. The addition of the new variables adds to the complexity of the problem's formulation, as it yields a set of m

equations in $m + n$ variables. To simplify this cumbersome notation, it has been suggested to use the simpler form $Ax = b$, with x containing in this case both the original and slack variables. As for A, it will contain, apart from the technological matrix, an additional unit matrix of order m. Hence, there is no need for x to be of dimension $n + m$. Rather, it shall be assigned a dimension n with the condition that n be larger than m. We can even suppose that the number of cost coefficients is n, some values of m being obviously null. Then, the problem can generally be stated as

$$\min_x \sum_{j=1}^{n} c_j x_j$$

subject to the condition that

$$x_j \geqq 0 \quad \text{given} \quad j = 1, \ldots, n$$

and

$$\sum_{j=1}^{n} a_{ij} x_j = b_i; \qquad i = 1, \ldots, m$$

The last equation can be rewritten in the following — more easily manageable — form, namely

$$x_1 p_1 + x_2 p_2 + \ldots + x_n p_n = p_0$$

where $p_1, \ldots, p_n$ are the columns of A, and $p_0 = b$ initially. In the sequel, we will suppose that $r(A) = m$, i.e. that A has at least m linearly independent vectors.

We come now to the two basic theorems of linear programming. Their proof, which will not be discussed in this text, is based on the concept of convexity, which is a master key to all results in the area. The two theorems can be stated as follows:

1. The objective function $F = \sum_{j=1}^{n} c_j x_j$ assumes its minimum only at an extreme point, i.e. at a vertex of the region X.
2. If a set of vectors $p_1, p_2, \ldots, p_k$, with $k \leqq m$, can be found such that they be linearly independent and, in addition

   $$x_1 p_1 + \ldots + x_k p_k = p_0, \; x_j \geqq 0$$

   then $x = (x_1, \ldots, x_k, 0, \ldots, 0)^T$ is an extreme point and is called a *basic feasible solution.*

These two theorems are the only basic tools really needed to solve the linear programming problem. There remain two further vital issues however, namely: how can one move from one basic feasible solution to another, and, second, checking at each time the value of the objective function F for every basic feasible solution to compare it with the value of F obtained just before. Although it can be done, one doesn't generally go through with such a procedure; a very tedious one especially for large problems, as there are in general $\binom{n}{m}$ vertices. Instead, one should resort to a scheme ensuring that one always moves from one vertex to another that has a smaller value of F.

Back to the first of the two issues, we suppose that there exists a basic feasible solution $(x_1, \dots, x_m, 0, \dots, 0)$: this is always the case since one can at least choose the slacks $x_i = b_i$, $i = 1, \dots, m$, the remaining $n - m$ variables being set equal to zero. Thence, we can write

$$x_1 p_1 + x_2 p_2 + \dots + x_m p_m = p_0 , \qquad x_j \geqq 0$$

Then moving from this vertex to another is equivalent to replacing p_1 for instance by p_{m+1} while making sure that $x_j \geqq 0$; $j = 2, \dots, m+1$. This is achieved by first expanding p_{m+1} in terms of $p_1, \dots, p_m$ i.e.

$$x_{1,m+1} p_1 + x_{2,m+1} p_2 + \dots + x_{m,m+1} p_m = p_{m+1}$$

Multiplying the expression by a factor θ and subtracting from the original form, we get

$$(x_1 - \theta x_{1,m+1})\, p_1 + \dots + (x_m - \theta x_{m,m+1})\, p_m + \theta p_{m+1} = p_0$$

In order for $x_1 - \theta x_{1,m+1}$, $x_2 - \theta x_{2,m+1}$, ... to be a basic feasible solution, one of the terms must vanish, the rest being positive with their corresponding vectors linearly independent. This is guaranteed by setting

$$0 \le \theta \le \min_j \frac{x_j}{x_{j,m+1}}$$

As for choosing among the column vectors of A a new basis that will guarantee our obtaining a new basic feasible solution having a smaller value of F, we must obviously use F itself, writing

$$x_{1j} p_1 + x_{2j} p_2 + \dots + x_{mj} p_m = p_j , \qquad j = 1, \dots, n$$

and

$$x_{1j} c_1 + x_{2j} c_2 + \dots + x_{mj} c_m = z_j , \qquad j = 1, \dots, n$$

where z_j is the objective function corresponding to the basic feasible solution $x_{1j}, x_{2j}, \dots, x_{mj}$; $c_1, \dots, c_m$ being their cost coefficients. If, at any instant of our performing the simplex, the relation $z_j - c_j > 0$ holds, then a set of feasible solutions can be constructed such that $z < z_0$, z being the value of F corresponding to a particular member of the set of basic feasible solutions. This can be deduced from the following analysis similar to the one depicted above:

$$(x_{10} - \theta x_{1j})\, p_1 + (x_{20} - \theta x_{2j})\, p_2 + \dots + (x_{m0} - \theta x_{mj})\, p_m + \theta p_j = p_0$$

and

$$(x_{10} - \theta x_{1j})\, c_1 + (x_{20} - \theta x_{2j})\, c_2 + \dots + (x_{m0} - \theta x_{mj})\, c_m + \theta c_j$$
$$= z_0 - \theta(z_j - c_j)$$

where θc_j was added to both sides. $x_{10}, \dots, x_{m0}$ is an initial basic feasible solution. Then, from the assumption that for a fixed j, $z_j - c_j > 0$, we get

$$z = z_0 - \theta(z_j - c_j) < z_0$$

Therefore, starting with the vector p_j, for which $z_j - c_j$ is positive and maximum, becomes the best choice for a new basis.

To illustrate the simplex method, we suggest the following example

$$\min F = x_1 - 3x_2 + 2x_3$$

subject to the conditions

$$\begin{aligned} 3x_1 - x_2 + 2x_3 &\leqq 7 \\ 2x_1 - 4x_2 &\leqq 12 \\ -4x_1 + 3x_2 + 8x_3 &\leqq 10 \\ x_1, x_2, x_3 &\geqq 0 \end{aligned}$$

i	Basis	c	p_0	1	−3	2	0	0	0
				p_1	p_2	p_3	p_4	p_5	p_6
1	4	0	7	3	−1	2	1	0	0
2	5	0	12	2	−4	0	0	1	0
3	6	0	10	−4	3	8	0	0	1
4	$z_j - c_j$			−1	3	−2	0	0	0

First Tableau

i	Basis	c	p_0	1	−3	2	0	0	0
				p_1	p_2	p_3	p_4	p_5	p_6
1	4	0	$\frac{31}{3}$	$\frac{5}{3}$	0	$\frac{14}{3}$	1	0	$\frac{1}{3}$
2	5	0	$\frac{76}{3}$	$\frac{-10}{3}$	0	$\frac{32}{3}$	0	1	$\frac{4}{3}$
3	2	−3	$\frac{10}{3}$	$\frac{-4}{3}$	1	$\frac{8}{3}$	0	0	$\frac{1}{3}$
4	$z_j - c_j$		−10	3	0	−10	0	0	−1

Second Tableau

i	Basis	c	p_0	p_1 (1)	p_2 (−3)	p_3 (2)	p_4 (0)	p_5 (0)	p_6 (0)
1	1	1	$\frac{31}{5}$	1	0	$\frac{14}{5}$	$\frac{3}{5}$	0	$\frac{1}{5}$
2	5	0	46	0	0	20	2	1	2
3	2	−3	$\frac{58}{5}$	0	1	$\frac{32}{5}$	$\frac{4}{5}$	0	$\frac{3}{5}$
4	$z_j - c_j$		$\frac{-143}{5}$	0	0	$\frac{-92}{5}$	$\frac{-9}{5}$	0	$\frac{-4}{5}$

Final Tableau

Our initial basis was p_4, p_5, p_6; the first basic feasible solution is (0, 0, 0, 7, 12, 10). $z_0 = 0$. p_2 is selected since $z_j - c_j$ is equal to 3. $\theta = 10/3$ and hence p_6 is eliminated. The new basis becomes p_4, p_5, p_2 and the second solution is $(0, \frac{10}{3}, 0, \frac{31}{3}, \frac{76}{3}, 0)$. Finally p_1 is introduced, since $z_j - c_j = 3$, $\theta = 31/5$ and hence p_4 is also eliminated. The new basis is p_1, p_5, p_2 and the solution comes as $(\frac{31}{5}, \frac{58}{5}, 0, 0, 46, 0)$, being the final solution as all $z_j - c_j$ are negative. $F_{\min} = -\frac{143}{5}$.

With this we have given a brief résumé of the main aspects of the linear programming problem. For further relevant techniques, the reader may further consult the texts of Dantzig (1963) and Gass (1969).

We come now to the major topic of the present chapter, namely sensitivity analysis in linear programming. The need for such an analysis arises from the question: what if the technological matrix A undergoes a perturbation ΔA? Also, what if the cost coefficients are subject to uncertainties? Although, according to Rappaport (1967), the "what if" type of questions introduces the topic of sensitivity in its very broad meaning, we are only interested here in its mathematical implications, namely in how the possible changes or errors in the parameters affect model outputs. There are more than one reason for performing sensitivity analysis, according to Anderson et al. (1979): raw material price fluctuations; stock exchange rates variations; employee turnover; purchasing of new equipment and so on. Sensitivity analysis is indeed an invaluable tool for the decision maker, helping him cope with the uncertainty and risks that are introduced by post-design systems variations. For this reason, it constitutes a good portion of any managerial science course.

More than one trend exist in handling sensitivity analysis in the context of linear programming. One very old trend is that of parametric programming, in which the parameters are allowed to vary with time, thus giving the linear programming problem an aspect of dynamism. Another more recent approach is to compute some critical values of the parameters after which the optimal solution changes from one basic feasible solution to another. A third trend is to model the problem in a stochastic form accounting for any uncertainty in the data.

To visualize the effect of uncertainty in the data on model output, consider the simple numerical example illustrated hereunder in Fig. 1.

$$\max_{x_1, x_2} 2x_1 + 3x_2$$

subject to the conditions

$$x_1, x_2 \geqq 0$$
$$x_1 \leqq 5$$
$$x_2 \leqq 6$$
$$x_1 + x_2 \leqq 8$$

and whose solution comes easily as $x_1 = 2$, $x_2 = 6$.

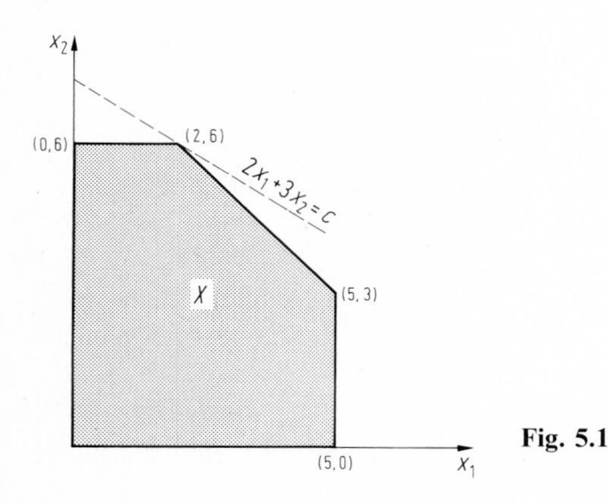

Fig. 5.1

Now suppose the problem is reformulated as

$$\max_{x_1, x_2} 2x_1 + 3x_2$$

subject to the conditions

$$x_1, x_2 \geqq 0$$
$$x_1 \leqq 5$$
$$x_2 \leqq 6 + \varepsilon$$
$$x_1 + x_2 \leqq 8$$

Then $x_{\text{optimal}} = (2 - \varepsilon, 6 + \varepsilon)^T$. A small change in the constraint has only slightly affected the solution. On the other hand, for the following problem

$$\max_{x_1, x_2} (2 + \varepsilon)\, x_1 + (3 - \varepsilon)\, x_2\,, \qquad \varepsilon > 0.5$$

subject to

$$x_1 \leqq 5$$
$$x_2 \leqq 6$$
$$x_1 + x_2 \leqq 8$$

x_{optimal} becomes $(5, 3)^T$. Thus a jump has occured from one vertex to another. In the first case, the initial basic feasible solution was maintained, though with a small perturbation. In the second instance, on the contrary, a complete shift of basis has taken place.

In this chapter, we shall assume that errors are solely due to uncertainties in the data; computational errors will be ignored. The process is then called *gutartig* — according to Meinguet (1969) — that is the errors produced are independent of the mode of computation. This definition appears in Bauer et al. (1965) as well, implying that the overall susceptibility of the solution to perturbations in the data dominates the numerical instability resulting from rounding errors.

5.2 Parametric Programming and Sensitivity Analysis

Unlike the case of linear simultaneous equations, sensitivity analysis in linear programming is rather involved. The matrix A in the system $Ax = b$ is of dimension $m \times n$, n being larger than m, and only m columns of A constitute a basic set at one time. Hence, not all perturbations in a_{ij}, with i and j varying from 1 to m and from 1 to n respectively, affect the optimal solution. Only those which form the entries of the base vectors are to be taken seriously, provided of course no change of basis has taken place. We have therefore two situations: one in which the perturbations in either A, b or c preserve the basis with only minor alterations in the solution; the second brings about noticeable change in the solution. These two problems have been studied in parallel in the literature, and are in fact interrelated. In this section, we will endeavour to investigate the first of the two problems.

The term *parametric programming* has originated in the context of finding the range of values of a parameter t in $a_{ij}(t)$ such that the basic solution remains the same. The earlier work on the subject is due to Saaty and Gass (1954), and Gass and Saaty (1955). The next section will indeed delve into their work. For the present, we can say that, by assuming that t is within range, or that it is allowed to vary while the basis is kept constant, one can study the effect of t in $a_{ij}(t)$, and consequently on $x(t)$. This has been studied in detail by Saaty (1959), the author reaching the first result in the subject of sensitivity analysis.

The linear programming problem can be formulated in terms of some parameter t as follows

$$\min_x \sum_{j=1}^{n} c_j(t)\, x_j$$

subject to

$$\sum_{j=1}^{n} a_{ij}(t)\, x_j \geq b_i\,, \qquad i = 1, \ldots, m$$

and

$$x_j \geqq 0\,, \qquad j = 1, \ldots, n\,.$$

Although in the author's original paper, the parameter t stood for time, it may as well serve as a perturbation factor like ε for instance; this is especially true as $a_{ij}(t)$ is usually given as a linear relation in t. In either case, our task is to obtain dx_j°/dt, thus eliciting the solution's sensitivity to changes in the data.

Let us first discuss the very same example chosen by Saaty (1959). The author considered

$$\min\ (30 + 6t)\, x_1 + (50 + 7t)\, x_2$$

subject to

$$(14 + 2t)\, x_1 + (4 + t)\, x_2 \leqq 14 + t$$
$$(150 + 3t)\, x_1 + (200 + 4t)\, x_2 \geqq 200 + 2t$$

and

$$x_1, x_2 \geqq 0$$

The results were

The solution vector $x^\circ(t) = x_1^\circ(t), \ldots, x_4^\circ(t)$	Intervals of the parameter t
$0\,, \dfrac{2t + 200}{4t + 200}\,, \dfrac{2t^2 + 48t + 2000}{4t + 200}\,, 0$	$t < -50\,, \quad t \geqq 10$
$\dfrac{2t^2 + 48t + 2000}{5t^2 + 294t + 2200}\,, \dfrac{t^2 + 236t + 700}{5t^2 + 294t + 2200}\,, 0\,, 0$	$-50 \leqq t < -49.9\,, \quad -6.1 \leqq t \leqq 10$
$0\,, \dfrac{t + 14}{t + 4}\,, 0\,, \dfrac{2t^2 + 48t + 2000}{t + 4}$	$-49.9 \leqq t \leqq -6.1$

The above results were obtained after tedious manipulations. Each time a set of base bectors had to be chosen, with the corresponding intervals of t calculated. This is in fact highly impractical, especially on the computer. Let us use it anyway to obtain some sensitivity calculations. For instance, if we are interested in $dx_j^\circ/dt; j = 1, 2, 3, 4$; at $t = 0$, we obtain

$$\left.\frac{dx_1^0}{dt}\right|_{t=0} = \left.\frac{(4t + 48)(5t^2 + 294t + 2200) - (2t^2 + 48t + 2000)(10t + 294)}{(5t^2 + 294t + 2200)^2}\right|_{t=0}$$

$$= -\frac{1206}{11 \times 1100}$$

$$\left.\frac{dx_2^0}{dt}\right|_{t=0} = \left.\frac{(2t + 236)(5t^2 + 294t + 2200) - (t^2 + 236t + 700)(10t + 294)}{(5t^2 + 294t + 2200)^2}\right|_{t=0}$$

$$= \frac{783 \cdot 5}{11 \times 1100}$$

and

$$\frac{dx_3^\circ}{dt} = \frac{dx_4^\circ}{dt} = 0$$

The above results can be more easily obtained if we know the base vectors. dx_j^0/dt, $j = 1, \dots, 4$, are calculated by differentiating with respect to t the equations $B(t)\, x^\circ = b(t)$, where B is a square matrix the columns of which are the base vectors. We obtain

$$\left.\frac{dx^0}{dt}\right|_{t=0} = B^{-1}(t)\left.\left\{\frac{db(t)}{dt} - \frac{dB(t)}{dt}\, x^0(t)\right\}\right|_{t=0}$$

$$= \begin{bmatrix} -14 & -4 \\ 150 & 200 \end{bmatrix}^{-1} \left\{ \begin{bmatrix} -1 \\ 2 \end{bmatrix} - \begin{bmatrix} -2 & -1 \\ 3 & 4 \end{bmatrix} \begin{bmatrix} \frac{10}{11} \\ \frac{7}{22} \end{bmatrix} \right\} = \begin{bmatrix} -\frac{1206}{11 \times 1100} \\ \frac{783 \cdot 5}{11 \times 1100} \end{bmatrix}$$

As for the sensitivity of the objective function F° with respect to t, we get

$$\left.\frac{dF^0}{dt}\right|_{t=0} = \frac{dc^T}{dt}\, x^0 + c^T \left.\frac{dx^0}{dt}\right|_{t=0}$$

$$= (6 \quad 7) \begin{bmatrix} \frac{10}{11} \\ \frac{7}{22} \end{bmatrix} + (30 \quad 50) \begin{bmatrix} -\frac{1206}{11 \times 1100} \\ \frac{783.5}{11 \times 1100} \end{bmatrix} = \frac{19189}{22 \times 110}$$

In fact, Saaty's main contribution lies in the second term above, namely $c^T(t)\,\dfrac{dx^\circ}{dt}$. The author discovered that this latter term can be written as

$$c^T \frac{dx^0}{dt} = c^T(t)\, B^{-1}(t) \left\{\frac{db(t)}{dt} - \frac{dB(t)}{dt}\, x^0(t)\right\} = y^{0^T}(t)\, \frac{db(t)}{dt} - y^{0^T}(t)\, \frac{dB(t)}{dt}\, x^0(t)$$

where $y^{0^T}(t) = c^T(t)\,B^{-1}(t)$ is the optimal solution of the *dual problem* to the original linear programming problem, i.e.; if the primary problem reads

$$\min_x \; c^T x$$

subject to the conditions

$$x \geqq 0\,, \qquad Ax \geqq b$$

the dual problem is (see Gass (1969)):

$$\max \; b^T y$$

subject to the conditions

$$y \geqq 0\,, \qquad A^T y \leqq c$$

And for Saaty's numerical example, the dual problem becomes

$$\max_{y_1, y_2} \; -(14 + t)\,y_1(t) + (200 + 2t)\,y_2(t)$$

subject to the conditions

$$\begin{aligned} &y_1, y_2 \geqq 0 \\ &-(14 + 2t)\,y_1(t) + (150 + 3t)\,y_2(t) \leqq 30 + 6t \\ &-(4 + t)\,y_1(t) + (200 + 4t)\,y_2(t) \leqq 50 + 7t \end{aligned}$$

Therefore, to obtain $\left.\dfrac{dF^0}{dt}\right|_{t=0}$ using Saaty's expression is equivalent to solving first the dual problem. The reader can follow the same steps of the simplex algorithm as explained in Sect. 4.1 to obtain $y_1^\circ(0)$ and $y_2^\circ(0)$. These were found to be

$$y_1^\circ = \frac{30}{44}\,, \qquad y_2^\circ = \frac{58}{5 \times 44}$$

Hence

$$\left.\frac{dF^0}{dt}\right|_{t=0} = (6 \;\; 7)\begin{bmatrix} \frac{10}{11} \\ \frac{7}{22} \end{bmatrix} + \left(\frac{30}{44} \;\; \frac{58}{44 \times 5}\right)\left\{\begin{bmatrix} -1 \\ 2 \end{bmatrix} - \begin{bmatrix} -2 & -1 \\ 3 & 4 \end{bmatrix}\begin{bmatrix} \frac{10}{11} \\ \frac{7}{22} \end{bmatrix}\right\}$$

$$= \frac{19\,189}{22 \times 110}$$

This result is consistent with the one obtained before. The final expression for dF°/dt thus reads

$$\frac{dF^0}{dt} = \frac{dc^T(t)}{dt}\, x^0(t) + y^{0^T}(t)\,\frac{db(t)}{dt} - y^{0^T}(t)\,\frac{dB(t)}{dt}\, x^0(t)$$

The application of the above result does not necessitate a formulation of the linear programming model in a parametric form. In fact, it becomes even more powerful in the usual case where the vectors b and c are constant. In such case, t represents a perturbation factor related to A, or, in other words, the differentiation can be imposed for each element a_{ij}. Thence we obtain the well known formula

$$\frac{\partial F^0}{\partial a_{ij}} = -y^{0T}\frac{\partial B}{\partial a_{ij}} x^0$$

$$= -y_i^0 x_j^0$$

which is a universal formula for the sensitivity analysis of a linear system of equations similar to the one derived in Sect. 1.6. Likewise, we could obtain the effect of an alteration of the vector b on the objective function, namely in the form

$$\frac{\partial F^\circ}{\partial b_i} = y_i^\circ$$

Webb (1962) extended Saaty's results to determine the most significant parameter having dominant influence over the optimal solution, and consequently over the objective function. It can be easily shown that

$$\frac{\partial x_k^\circ}{\partial a_{ij}} = -y_i^\circ x_j^\circ / c_k$$

For results similar to those of Saaty in the cases where F is convex quadratic function in x, the reader is referred to Boot (1963). Some other relevant applications are also to be found in Van De Panne and Bosje (1962).

Another approach to assessing the sensitivity of x° with respect to changes in the parameters of the problem $Bx^\circ = b$ is to obtain an a-priori bound like the one obtained in Sect. 1.4. Here, it would be given by

$$\|\Delta x^\circ\| \leqq \|B^{-1}\|\ \|\Delta B x^\circ - \Delta b\|$$

where x° is the optimal solution of the unperturbed problem. $\|B^{-1}\|$ can be estimated from

$$\|B^{-1}\| = \max \|y\|\,, \qquad By = z\,, \qquad \|z\| = 1$$

(see Sect. 1.3). For a general linear programming problem composed of set of equalities and inequalities, i.e.

$$\max c^T x$$

subject to the restrictions $Ax \leqq b$, $Dx = d$, one has from Robinson (1973)

$$\|\Delta x^0\| \leq \sigma(A, D) \left\| \begin{matrix} \Delta A x^0 - \Delta b \\ \Delta D x^0 - \Delta d \end{matrix} \right\|$$

where

$$\sigma(A, D) = \max \{\min \{\|x\|, Ax \leqq b, Dx = d\} \ \|b, d\| \leqq 1\}$$

The above bound follows from a result of Hoffman's (1952) on linear inequalities. For further extensions on perturbations of a set described by linear inequalities, the reader is referred to Daniel (1973).

To exemplify Saaty's results, let us consider

$$\min 30x_1 + 50x_2$$

subject to the conditions

$$x_1, x_2 \geqq 0$$
$$14(1 + \varepsilon)\, x_1 + 4x_2 \leqq 14$$
$$150x_1 + 200x_2 \geqq 200$$

By inducing a perturbation ε in a_{11}, we obtain the new objective function, for small enough values of $|\varepsilon|$, which comes as

$$\begin{aligned} F^\circ(a_{11} + \Delta a_{11}) &\approx F^\circ(a_{11}) + \Delta a_{11} \frac{\partial F^\circ}{\partial a_{11}} \\ &= \frac{475}{11} + (-14\varepsilon)(-y_1^\circ \cdot x_1^\circ) \\ &= \frac{475}{11} + 14\varepsilon \cdot \frac{30}{44} \cdot \frac{10}{11} \\ &= \frac{475}{11} + \frac{1050}{121}\varepsilon \end{aligned}$$

This result coincides with the one ultimately obtained after a number of cumbersome manipulations and depicted in the following final tableau:

Basis	c	p_0	30	50	0	0
			p_1	p_2	p_3	p_4
1	30	$\frac{10}{11 + 14\varepsilon}$	1	0	$\frac{1}{11 + 14\varepsilon}$	$\frac{2}{1100 + 1400\varepsilon}$
2	50	$\frac{7 + 28\varepsilon}{22 + 28\varepsilon}$	0	1	$\frac{-15}{220 + 280\varepsilon}$	$\frac{-7(1 + \varepsilon)}{1100 + 1400\varepsilon}$
$z_j - c_j$		F°	0	0	$\frac{-15}{22 + 28\varepsilon}$	$\frac{-29}{110 + 140\varepsilon}$

with

$$F^\circ = \frac{950 + 1400\varepsilon}{22 + 28\varepsilon} \cong \frac{475}{11} + \frac{1050}{121}\varepsilon .$$

Note that, in the tableau, we have allowed a small perturbation in a_{11} of order ε that will still maintain the same basis. This is guaranteed by choosing ε subject to the restriction

$$22 + 28\varepsilon > 0$$

or that

$$\varepsilon > -\frac{11}{14}$$

On the other hand, if ε violates the value set above, a totally different sequence of events will be triggered as will be seen later on.

In conclusion, we can say that parametric programming, while being difficult to use on a computer, serves as yet to compute sensitivites at different levels of ε or t, provided the perturbations in the data are of small order. It's also invaluable when it comes to establishing a correspondence between solution and interval of parameter values, a fact that we will try to explain in the coming section.

5.3 The Problem of Ranging in Linear Programming

In many a practical situation, it is rather often desirable to determine that range of the parameters — or data — that will not alter the solution significantly. This problem is of vital importance in many areas of decision-making and management sciences, for, whereas data for an analyst are merely terms of a set of linear equations, these same data embrace a definite practical reality for the executive. As Sandor (1964) pointed out, these data may be sales figures of home heating fuel, liable to fluctuation with climatic variations, or the purchase prices of this same commodity, which are subject to frequent variation. A primordial question would then be: which ranges of data could be tolerated for the same optimal solution to remain valid? In management science, this question is synonymous to seeking a definition of the *decision region* for each of the parameters.

This problem was dealt with within the context of parametric programming by Saaty and Gass (1954) and Gass and Saaty (1955), for the case where the data is given in parametric form. A more efficient approach is given by Shetty (1959a, 1959b), relying only on the final simplex tableau.

Saaty and Gass (1954) suggested a linear programming problem of the form

$$\min_x \sum_{j=1}^{n} c_j(\lambda)\, x_j$$

subject to the conditions

$$\sum_{j=1}^{n} a_{ij}(\lambda)\, x_j \geq b_i\,, \qquad i = 1, \ldots, m$$

and

$$x_j \geqq 0\,, \qquad j = 1, \ldots, n$$

The authors' main concern was to locate the optimal solution x_j, $j = 1, \ldots, n$, corresponding to a certain range of the parameter λ, or, in other words, for which range of interval of values of λ will the optimal solution remain optimal? This necessitates going through all the steps of the simplex in terms of λ and, at each time we locate a minimum, the corresponding values of λ are checked out. Therefore, to each minimum solution, there exists one interval of values of λ that is generally closed (nondegeneracy assumed). Each new interval is contiguous with the previous one, meaning that they meet at end points.

To determine the interval of values of λ corresponding to a minimum, we observe $z_j - c_j \leqq 0$, for all values of j. The latter will be a function of λ in the contracted form $\alpha_j + \lambda\beta_j$, assuming that the data is expressed as a series in λ. Therefore, if a minimum is found for $\lambda = \lambda_0$, then

$$\alpha_j + \lambda_0\beta_j \leqq 0\,, \qquad j = 1, \ldots, n$$

determines that region of λ_0 for which the above minimum remains an optimum solution. The minimal solution for $\lambda = \lambda_0$ is then a minimum for all values of λ, such that

$$\max_{\beta_j > 0} (-\alpha_j/\beta_j) \leq \lambda \leq \min_{\beta_j > 0} (-\alpha_j/\beta_j)$$

A typical example has been studied by Saaty (1959), demonstrating the above procedure, its results being tabulated in Sect. 4.2. The two authors elaborated on the subject still further in their work (Gass and Saaty (1955)) by introducing the two-parameter problem

$$\min_{x} \sum_{j=1}^{n} (a_j + \lambda_1 b_j + \lambda_2 c_j)\, x_j$$

subject to the conditions $x_j \geqq 0$, $j = 1, \ldots, n$ and

$$\sum_{j=1}^{n} a_{ij} x_j = a_{i0}\,, \qquad i = 1, \ldots, m$$

The minimum solution is the one satisfying the set of inequalities

$$z_j - (a_j + \lambda_1 b_j + \lambda_2 c_j) \leqq 0\,, \qquad \forall j = 1, \ldots, n$$

which define a convex region in the (λ_1, λ_2) plane termed, according to the authors, the *characteristic region*, the boundaries of which being accordingly termed the *characteristic boundaries*.

To visualize this approach, we consider the same example presented by the authors:

$$\min_{x_1, x_2, x_3} x_1 + \lambda_1 x_2 + \lambda_2 x_3$$

such that

$$\begin{aligned} &x_1, x_2, x_3 \geqq 0 \\ &x_1 - x_4 - 2x_6 = 5 \\ &x_2 + 2x_4 - 3x_5 + x_6 = 3 \\ &x_3 + 2x_4 - 5x_5 + 6x_6 = 5 \end{aligned}$$

The tableau is

Basis	c	p_0	1	λ_1	λ_2	0	0	0
			p_1	p_2	p_3	p_4	p_5	p_6
1	1	5	1	0	0	-1	0	-2
2	λ_1	3	0	1	0	2	-3	1
3	λ_2	5	0	0	1	2	-5	6
$F_j = z_j - c_j$		F_0	F_1	F_2	F_3	F_4	F_5	F_6

where

$$\begin{aligned} F_0 &= 5 + 3\lambda_1 + 5\lambda_2 \\ F_1 &= F_2 = F_3 = 0 \\ F_4 &= -1 + 2\lambda_1 + 2\lambda_2 \\ F_5 &= -3\lambda_1 - 5\lambda_2 \\ F_6 &= -2 + \lambda_1 + 6\lambda_2 \end{aligned}$$

Hence for the vectors p_1, p_2 and p_3 to remain base vectors, one must have

$$\begin{aligned} &\lambda_1 + \lambda_2 \leqq \tfrac{1}{2} \\ &-3\lambda_1 - 5\lambda_2 \leqq 0 \\ &\lambda_1 + 6\lambda_2 \leqq 2 \end{aligned}$$

conditions which define the region C_1 depicted in Fig. 2. Then by changing the basis, we obtain the other unbounded regions as shown.

Note that, whenever λ lies inside one region, then a sensitivity analysis, as explained in Sect. 5.2, can be carried out so long as the perturbation in λ does not

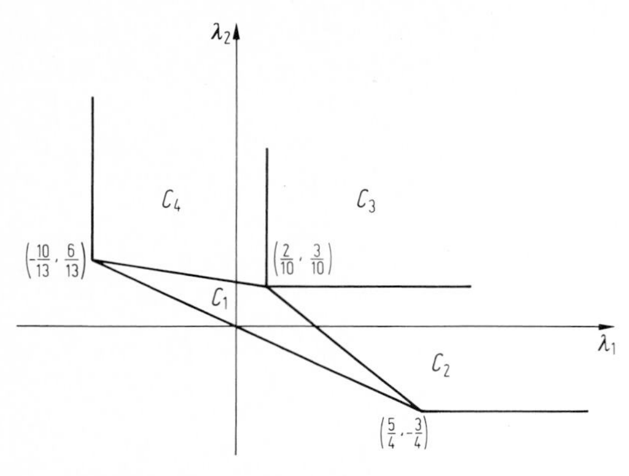

Fig. 5.2

overlap with another region. The objective function F° becomes, for small perturbations around $\lambda_0 \in C_1$:

$$F^0(\lambda) \approx F^0(\lambda_0) + \left.\frac{dF^0}{d\lambda}\right|_{\lambda_0} \Delta\lambda$$

On the other hand, when $\Delta\lambda$ is large, an exact representation is more appropriate, namely

$$\begin{aligned} F^\circ(\lambda) &= C^T(\lambda)\, x^\circ(\lambda) \\ &= C^T(\lambda_0 + \Delta\lambda)\,\{B(\lambda_0) + \Delta\lambda B_1\}^{-1}\,\{b(\lambda_0) + \Delta\lambda b_1\} \end{aligned}$$

where $B(\lambda_0)$ is the basis matrix at λ_0. The inverse of $\{B(\lambda_0) + \Delta\lambda B_1\}$ is computed as a series in $\Delta\lambda$ as seen in Sect. 1.4. For the linear programming problem, however, as perturbations are introduced to each parameter one at a time, the above inverse can be easily given in an explicit form (see exercise 5.1).

In Saaty and Gass's work, the data is assumed to be parameter-dependent, and the authors exercised in finding the correspondence between optimal solutions and ranges of parameter values. In practice, one has rarely data in a parametric form. Furthermore, having the data in this form would have complicated things seriously, especially for large sized models, where the use of a computer becomes unavoidable. This has inspired Shetty (1959a, 1959b) to devise a set of algorithms that systematically set out the correspondence between feasible solutions and the ranges of data available. The advantage of Shetty's scheme is all the more accentuated by the fact that it only requires the results of the final tableau. Thence, surprinsingly enough, although the original problem might have had to be inescapably solved on the computer, one would only need manual calculations to deduce all sensitivity formulae. Shetty specifically addressed the following two questions:

1. What becomes of the optimal solution when one of the problem's constants (some c_j, b_i or a_{ij}) is changed without yet necessitating a change of basis? What is the new value of the objective function?
2. When one of the variables' value is incremented by a given amount, what changes become necessary in the other variables' values if the increase in the value of $F(x)$ is to be minimal.

Although the first question seemingly repeats what was said in Sect. 5.2 regarding solution sensitivity with respect to changes in the data, Shetty's answering it provides in fact an answer to the extended problem of intervals of values of data that preserve basis. In the sequel, we shall use the same notation as in Shetty's papers, and use his example once we have transformed it into a minimization problem for the sake of homogeneity in our text. In this revised form, the example comes as

$$\min -20x_1 - 10x_2 - 20x_3 + 0x_4 + 0x_5 + 0x_6$$

subject to the conditions

$$x_1, x_2, x_3, x_4, x_5, x_6 \geqq 0$$
$$3x_1 + x_2 + 4x_3 + x_4 = 480$$
$$4x_1 + 2x_2 + 3x_3 + x_5 = 400$$
$$x_2 + x_6 = 70$$

where x_4, x_5, x_6 are slacks. The initial tableau is

Basis	c	p_0	-20	-10	-20	0	0	0
			p_1	p_2	p_3	p_4	p_5	p_6
4	0	480	3	1	4	1	0	0
5	0	400	4	2	3	0	1	0
6	0	70	0	1	0	0	0	1
$w_j = z_j - c_j$		0	20	10	20	0	0	0

The final tableau is

Basis	c	p_0	-20	-10	-20	0	0	0
			p_1	p_2	p_3	p_4	p_5	p_6
3	-20	112	0.4	0	1	0.4	-0.2	0
2	-10	32	1.4	1	0	-0.6	0.8	0
6	0	38	-1.4	0	0	0.6	-0.8	1
	z'_j	-2560	-22	-10	-20	-2	-4	0
$w'_j = z'_j - c_j =$			-2	0	0	-2	-4	0

As for the notation, like in the Shetty's paper, we shall assume that B is the matrix containing the basis, i.e. $B = \{j \mid x_j \in \text{basis in the initial tableau}\}$, whilst $B' = \{j \mid x_j \in \text{basis in the final tableau}\}$. The element at the intersection of the row corresponding to a variable x_p $(p \in B)$ and the column corresponding to a variable x_q will be denoted by x_{pq}. Likewise, b_p will lie in the x_p's row. For example, $a_{52} = 2$, $b_4 = 480$, $w_2 = 10$, $a'_{31} = 0.4$, $b'_3 = 112$, $z'_1 = -22$ and $w'_1 = -2$. We will now investigate Shetty's algorithms:

Variation in c_j

Case I: $j \notin B'$, i.e. x_j is not in the basis of the final tableau.

a) The maximum possible change in c_j without changing the basis in the final tableau is given by

$$w'_j \leqq \Delta c_j \leqq \infty\,, \qquad j \notin B'$$

This follows from

$$w'_j + \Delta w'_j \leqq 0\,, \qquad j = 1, \ldots, n$$

b) The solution x, as well as the objective function F, is unaffected, that is:

$$\Delta x = 0\,, \qquad \Delta F = 0$$

Case II: $j \in B'$, i.e. x_j is basis variable in the final tableau.

a) The maximum possible change in c_j without changing the basis is given by

$$\underline{\Delta c}_j \leqq \Delta c_j \leqq \overline{\Delta c}_j$$

where

$$\underline{\Delta c}_j = \max\left[\frac{-w'_k}{a'_{jk}}\,, -\infty\right], \qquad a'_{jk} < 0\,, \qquad k \notin B'$$

$$\overline{\Delta c}_j = \min\left[\frac{-w'_k}{a'_{jk}}\,, \infty\right], \qquad a'_{jk} > 0\,, \qquad k \notin B'$$

This follows from

$$w'_k + \Delta w'_k \leqq 0\,, \qquad k = 1, \ldots, n$$

$$\Delta w'_k = a'_{jk}\,\Delta c_j$$

b) The solution x is unaffected, i.e. $\Delta x = 0$, while

$$\Delta F = \Delta c_j x_j$$

Example: In the above illustrative example, if c_1 is decreased by more than 2 units, the second tableau will consequently not be the final one. Also, if c_2 is increased by more than the following

$$\overline{\Delta c}_2 = \min\left[\frac{2}{1.4}\,, \frac{4}{0.8}\right] = 1.43$$

table 2 will again not be the final tableau. Similarly, the maximum reduction in c_2 is

$$\underline{\Delta}c_2 = \frac{2}{-0.6} = -3.33$$

Variation in b_i

Suppose b_i is changed within the interval

$$\Delta b_i \in [\underline{\Delta}b_i, \bar{\Delta}b_i],$$

then to keep the basis, one must have:

$$\underline{\Delta}b_i = \max\left[\frac{-x_j}{a'_{ji}}, -\infty\right], \qquad a'_{ji} > 0, \qquad j \in B'$$

$$\bar{\Delta}b_i = \min\left[\frac{-x_j}{a'_{ji}}, \infty\right], \qquad a'_{ji} < 0 \qquad j \in B'$$

This follows from

$$x_j + \Delta x_j \geqq 0, \qquad j \in B'.$$

And as

$$x + \Delta x = B^{-1}(b + \Delta b),$$

we get

$$x_j + a'_{ji}\,\Delta b_i \geqq 0, \qquad j \in B'$$

from which the above follows.

Now, for the change in x_j, we directly have

$$\Delta x_j = \begin{cases} \Delta b_i a'_{ji}, & j \in B' \\ 0, & j \notin B' \end{cases}$$

The change in F is given by

$$\Delta F = \sum_{j \in B'} c_j\,\Delta x_j = \sum_{j \in B'} c_j a'_{ji}\,\Delta b_i$$

$$= \Delta b_i z'_i$$

Example: In The above illustrative numerical example,

$$\bar{\Delta}b_4 = \frac{-32}{-0.6} = 53.33$$

$$\underline{\Delta}b_4 = \max\left[\frac{-112}{0.4}, \frac{-38}{0.6}\right] = -63.33$$

Variation in a_{ij}

Case I: $j \in B'$, i.e. x_j is a basis variable in the final tableau.

a) The maximum possible change in a_{ij} without necessitating a change of the basis in the final tableau is given by

$$\overline{\Delta}a_{ij} = \min\left[\frac{x_k}{(a'_{ki}x_j - a'_{ji}x_k)}, \frac{w'_l}{(z'_i a'_{jl} - w'_l a'_{ji})}, \infty\right]$$

$$\underline{\Delta}a_{ij} = \max\left[\frac{x_k}{(a'_{ki}x_j - a'_{ji}x_k)}, \frac{w'_l}{(z'_i a'_{jl} - w'_l a'_{ji})}, -\infty\right]$$

where the first term in $\overline{\Delta}a_{ij}$ is evaluated for all $k \in B'$, satisfying the relation $a'_{ki}x_j > \max\,[0, a'_{ji}x_k]$ while the second term for all $l \notin B'$ satisfies the relation $z'_i a'_{jl} < \min\,[0, w'_l a'_{ji}]$. Likewise, for $\underline{\Delta}a_{ij}$, the inequality is reversed and the maximum is replaced by a minimum.

b) For a change in a_{ij} satisfying the relation

$$\underline{\Delta}a_{ij} \leqq \Delta a_{ij} \leqq \overline{\Delta}a_{ij},$$

the optimal change required in the values of the variables is given by

$$\Delta x_k = \begin{cases} \dfrac{-\Delta a_{ij} a'_{ki} x_j}{1 + \Delta a_{ij} a'_{ji}}, & k \in B' \\ 0, & k \notin B' \end{cases}$$

c) The change in the value of F is given by

$$\Delta F = \frac{-\Delta a_{ij} z'_i x_j}{1 + \Delta a_{ij} a'_{ji}}$$

To prove the points a, b and c, we write the matrix $\overline{B}$ of the base vectors, subject to perturbations Δa_{ij} in the element a_{ij}, in the form

$$\begin{aligned}\overline{B} &= B + \Delta a_{ij} E \\ &= B(I + B^{-1}\,\Delta a_{ij} E)\end{aligned}$$

where E is an $m \times m$ matrix with $e_{ij} = 1$ and zero otherwise. Hence

$$\begin{aligned}\overline{B}^{-1} &= (I + B^{-1}\,\Delta a_{ij} E)^{-1}\,B^{-1} \\ &= B^{-1} - \frac{\Delta a_{ij}}{1 + \Delta a_{ij} a'_{ji}}\,B^{-1} E\, B^{-1}\end{aligned}$$

This last expression is a substitute for the expansion in Sect. 1.4, in the case of perturbation in one element only (see exercise 5.1). And as

$$\Delta x = (\overline{B}^{-1} - B^{-1})\,b,$$

it follows that

$$\Delta x_k = \frac{-\Delta a_{ij} a'_{ki} x_j}{1 + \Delta a_{ij} a'_{ji}}, \qquad k \in B'$$

which proves point *b*. To prove point *c*, write

$$\begin{aligned} \Delta F &= \sum_{k \in B'} c_k \, \Delta x_k \\ &= \frac{-\Delta a_{ij} z'_i x_j}{1 + \Delta a_{ij} a'_{ji}} \end{aligned}$$

Lastly, to prove point *a*, consider the change in w'_l as

$$\begin{aligned} \Delta w'_l &= \begin{cases} 0, & l \in B' \\ \sum_{k \in B'} c_k \, \Delta a'_{kl}, & l \notin B' \end{cases} \\ &= \sum_{k \in B'} c_k \left(\frac{-\Delta a_{ij} a'_{ki} a'_{jl}}{1 + \Delta a_{ij} a'_{ji}} \right), \qquad l \notin B' \\ &= \frac{-\Delta a_{ij} z'_i a'_{jl}}{1 + \Delta a_{ij} a'_{ji}}, \end{aligned}$$

from which point a follows, when we use

$$x_k + \Delta x_k \geqq 0, \qquad w'_l + \Delta w'_l \leqq 0$$

Case II: $j \notin B'$, i.e. x_j is not a basis variable in the final tableau.

a) The maximum possible change in a_{ij}, without necessitating a change in the basis is given by

$$\underline{\Delta a}_{ij} \leqq \Delta a_{ij} \leqq \overline{\Delta a}_{ij}$$

where

$$\underline{\Delta a}_{ij} = \begin{cases} -\infty, & z'_i \geq 0 \\ \dfrac{-w'_j}{z'_i}, & z'_i < 0 \end{cases}$$

$$\overline{\Delta a}_{ij} = \begin{cases} \infty, & z'_i \leq 0 \\ \dfrac{-w'_j}{z'_i}, & z'_i > 0 \end{cases}$$

b) For a change in a_{ij} within the above limits, we have

$$\Delta x_i = 0$$

i.e. the solution is optimal.

c) For a change in a_{ij} within the limits indicated above, the change in F is given by

$$\Delta F = 0$$

To prove assertion a, we consider

$$\Delta a'_{kj} = a'_{ki}\,\Delta a_{ij}$$

and since

$$\Delta w'_j = \sum_{k \in B'} c_k\,\Delta a'_{kj}$$

$$= \sum_{k \in B'} c_k a'_{ki}\,\Delta a_{ij}$$

$$= \Delta a_{ij} z'_i$$

the proof is complete if we take $w'_j + \Delta w'_j \leqq 0$. Assertions b and c are trivial.

Example: In the illustrative example above, suppose $\Delta a_{42} = -0.5$. From a in case I, there are no rows or columns satisfying the conditions for evaluating the first two terms, and we get $\underline{\Delta a}_{42} = -\infty$. As for the optimal change in x_k, $k \in B'$, we have

$$\Delta x_3 = \frac{-(-0.5)\,(0.4)\,(32)}{1 + (-0.5)\,(-0.6)} = 4.9$$

$$\Delta x_2 = \frac{-(-0.5)\,(-0.6)\,(32)}{1 + (-0.5)\,(-0.6)} = -7.4$$

$$\Delta x_6 = \frac{-(-0.5)\,(0.6)\,(32)}{1 + (-0.5)\,(-0.6)} = 7.4$$

Likewise, the change in F is given by

$$\Delta F = \frac{-(-0.5)\,(-2)\,(32)}{1 + (-0.5)\,(-0.6)} = -24.6$$

Suppose instead that we wish to determine the maximum change allowed in a_{41} to preserve the basis. Applying algorithm a in case II, we obtain

$$\underline{\Delta a}_{41} = -\frac{-2}{-2} = -1$$

with no change in x_3, x_2, x_6 or F.

Variation in x_j optimal

Suppose, after we have computed the optimal solution, a variable x_k is allowed to vary. The second question Shetty (1959a) asked was: what is the change required in the other x_j, $j = 1, \ldots, k, \ldots, n$, so that the increase in F be minimum?

Case I: $k \notin B'$, i.e. x_k is not a basis variable in the final tableau.

a) The maximum possible change in x_k, without changing the basis in the final tableau, is given by:

$$0 \leqq \Delta x_k \leqq \bar{\Delta} x_k$$

where

$$\bar{\Delta} x_k = \min\left[\frac{x_i}{a'_{ik}}, \infty\right], \qquad i \in B', \qquad a'_{ik} > 0 .$$

Note that $\underline{\Delta} x_k = 0$ since $x_k = 0$, $k \notin B'$

b) For a change in x_k within the limits given above, in order for the increase to be minimum, the optimal change in the values of the variables is given by:

$$\Delta x_i = \begin{cases} -a'_{ik}\, \Delta x_k, & i \in B' \\ 0, & i \notin B', \quad i \neq k \\ \Delta x_k, & i = k \end{cases}$$

c) The corresponding change in F is given by:

$$\Delta F = -w'_k\, \Delta x_k$$

To prove the above assertions, we first note that for the change in F to be minimum, one must have $\Delta b'_i = 0$. Hence, we have

$$\Delta x_i + a'_{ik}\, \Delta x_k = 0 ,$$

and assertion a follows, by imposing that $x_i + \Delta x_i \geqq 0$. Assertion *b* is trivial. As for assertion *c*, we have that

$$\Delta F = \sum_{i \in B'} c_i\, \Delta x_i = -\sum_{i \in B'} c_i a'_{ik}\, \Delta x_k$$

$$= -z'_k\, \Delta x_k = -(w'_k + c'_k)\, \Delta x_k = -w'_k\, \Delta x_k$$

Case II: $i \in B'$, i.e. x_i is a basis variable in the final tableau.

a) A change in x_i, $i \in B'$ will require a maximum possible change in x_I, $I \in B'$, such that

$$\underline{\Delta} x_I = \max\left[\frac{-a'_{Ik} x_i}{a'_{ik}}\right], \qquad i \in B', \qquad a'_{ik} > 0$$

$$\bar{\Delta} x_I = \min\left[\frac{-a'_{Ik} x_i}{a'_{ik}}\right], \qquad i \in B', \qquad a'_{ik} < 0$$

b) For a change in x_I within the limits indicated above, in order for the change in ΔF to be minimum, the change required in the values of the variables is given by:

$$\Delta x_i = \begin{cases} -\dfrac{\Delta x_I}{a'_{Ik}}, & i = k \\ 0, & i \notin B', \quad i \neq k \\ \dfrac{a'_{ik}\, \Delta x_I}{a'_{Ik}}, & i \in B' \end{cases}$$

c) The corresponding change in ΔF is given by:

$$\Delta F = \frac{w'_k\, \Delta x_I}{a'_{Ik}}$$

To prove the above assertions, we note that since F depends only on c_i, $i \in B'$, a change in x_i necessitates changes in x_I, $I \in B'$. Therefore, from

$$\Delta x_i + a'_{ik}\, \Delta x_k = 0, \qquad i \in B', \qquad k \notin B'$$

$$\Delta x_I + a'_{Ik}\, \Delta x_k = 0, \qquad I \in B', \qquad k \notin B'$$

we can deduce that

$$\Delta x_i - \Delta x_I \frac{a'_{ik}}{a'_{Ik}} = 0$$

And from $x_i + \Delta x_i \geqq 0$, assertion a follows, so does assertion b. As for assertion c, we have

$$\Delta F = \sum_{i \in B'} c_i \Delta x_i = \sum_{i \in B'} - c_i a'_{ik} \Delta x_k$$

$$= -w'_k \Delta x_k = w'_k \frac{\Delta x_I}{a'_{Ik}}$$

Example: suppose in the illustrative example above, that $\Delta x_1 = 10$, then

$$\Delta x_3 = -4$$
$$\Delta x_2 = -14$$
$$\Delta x_6 = 14$$

Also

$$\Delta F = -(-2)\,(10) = 20$$

i.e. F will increase by 20 units. On the other hand, if x_2 must be incremented by 18, we must then have

$$\Delta x_4 = \frac{-18}{-0.6} = 30$$

$$\Delta x_3 = \frac{0.4 \times 18}{-0.6} = -12$$

$$\Delta x_6 = \frac{0.6 \times 18}{-0.6} = -18$$

with a corresponding increase in F equal to

$$\Delta F = \frac{-2 \times 18}{-0.6} = 60$$

This concludes our discussion of the problem of ranging in linear programming. Similar ideas to those of Shetty (1959a, 1959b) can be found in Courtillot (1960) though in a less systematic form.

5.4 Interval Programming

An *interval linear programming* problem is defined as

$$\max_x F = c^T x$$

subject to the condition

$$b^- \leqq Ax \leqq b^+$$

A vector x satisfying the above inequalities is called a feasible solution to the linear programming problem. If max $c^T x$ is also finite, then x is optimal and the problem is termed bounded.

Many practical problems can be formulated in this fashion. For example, the problem discussed in Sect. 2.7 can be treated using interval linear programming. Many other problems fall under this same category as well. In the literature, we find applications to chemicals manufacturing, etc. ...

Although interval linear programming can be performed using solely the standard simplex, and this by increasing the number of inequalities, such a transformation will however significantly increase the effective size of the problem. Ben-Israël and Charnes (1968) were for their part able to express explicitly the optimal solution x in terms of A, b^+ and b^-. They also showed that F is bounded if c is orthogonal to $N(A)$, which is the nullity of A. To prove the authors' proposition, the inequalities $b^- \leqq Ax \leqq b^+$ are concisely written in the form

$$Ax = b\,, \qquad b \in [b^-, b^+]\,.$$

Hence it follows that

$$x = A^i b + u\,, \qquad b \in [b^-, b^+]\,, \qquad u \in N(A)\,,$$

where A^i is a generalized inverse of A satisfying $AA^iA = A$ (see Sect. 4.1). The above solution exists only if the equations are consistent, i.e.

$$AA^i b \subseteq [b^-, b^+]\,, \qquad b \in [b^-, b^+]$$

(cf. Sect. 2.6). Hence if A is of full row rank, i.e. if $r(A) = m$, then $AA^i = I$, and the above condition is guaranteed. This is the case investigated by Ben-Israël and Charnes (1968). Furthermore, the objective function F becomes

$$\begin{aligned} F &= c^T x \\ &= c^T A^i b + c^T u\,, \qquad b \in [b^-, b^+]\,, \qquad u \in N(A) \end{aligned}$$

Then it follows that, for F to be finite, one must have

$$c^T u = 0$$

which is the authors' first result. Next, to maximize F, one must choose among the extreme values of $[b^-, b^+]$ those which yield a maximum F when premultiplied by $c^T A^i$. This is achieved by observing the signs of $c^T a^i_j$, where a^i_j is the jth column of A^i; $j = 1, \ldots, m$. Hence by associating similar signs in $[b^-, b^+]$ and $c^T a^i_j$, one can maximise F. Likewise, associating opposite signs minimizes F. Therefore, for a maximum interval linear programming problem, we obtain the optimal solution x^0 in the form

$$x^0 = \sum_{j^-} a^i_j b^-_j + \sum_{j^+} a^i_j b^+_j + \sum_{j^0} a^i_j [\sigma_j b^+_j + (1 - \sigma_j)\, b^-_j] + u\,,$$

where $\sum_-$, $\sum_+$ and $\sum_0$ are summations performed over j on negative, positive and zero values of $c^T a^i_j$ respectively and $0 \leqq \sigma_j \leqq 1$.

The above results are illustrated in the following simple example

$$\max x\,, \qquad \max y\,, \qquad \min x\,, \qquad \min y$$

subject to the conditions

$$\begin{aligned} & x + y \geqq 5 \\ & x + y \leqq 10 \\ & y \geqq x \\ & y \leqq x + 3 \end{aligned}$$

i.e.

$$\begin{bmatrix} 5 \\ 0 \end{bmatrix} \leq \begin{bmatrix} 1 & 1 \\ -1 & 1 \end{bmatrix} \begin{bmatrix} x \\ y \end{bmatrix} \leq \begin{bmatrix} 10 \\ 3 \end{bmatrix}$$

and

$$A^i = \begin{bmatrix} \frac{1}{2} & -\frac{1}{2} \\ \frac{1}{2} & \frac{1}{2} \end{bmatrix}$$

To maximize x, we have that

$$F^0 = \max c^T x = \max [1 \quad 0] \begin{bmatrix} \frac{1}{2} & -\frac{1}{2} \\ \frac{1}{2} & \frac{1}{2} \end{bmatrix} \begin{bmatrix} [5, 10] \\ [0, 3] \end{bmatrix} = 5 + 0 = 5$$

As for the maximization of y, we have that

$$F^0 = \max c^T x = [0 \quad 1] \begin{bmatrix} \frac{1}{2} & -\frac{1}{2} \\ \frac{1}{2} & \frac{1}{2} \end{bmatrix} \begin{bmatrix} [5, 10] \\ [0, 3] \end{bmatrix} = 5 + \frac{3}{2} = \frac{13}{2}$$

Likewise, to minimize x, we have that

$$F^0 = \min c^T x = [1 \quad 0] \begin{bmatrix} \frac{1}{2} & -\frac{1}{2} \\ \frac{1}{2} & \frac{1}{2} \end{bmatrix} \begin{bmatrix} [5, 10] \\ [0, 3] \end{bmatrix} = \frac{5}{2} - \frac{3}{2} = 1$$

And lastly, to minimize y, we have that

$$F^0 = \min c^T x = [0 \quad 1] \begin{bmatrix} \frac{1}{2} & -\frac{1}{2} \\ \frac{1}{2} & \frac{1}{2} \end{bmatrix} \begin{bmatrix} [5, 10] \\ [0, 3] \end{bmatrix} = \frac{5}{2} + 0 = \frac{5}{2}$$

The above method is valid if A is of full row rank. If $r(A) < m$, then the consistency condition of the equations must be checked out first, to make sure that for all $b \in [b^-, b^+]$, one has

$$AA^i b \subseteq [b^-, b^+]$$

The problem becomes more complicated if $m > n$, for although x can be written in the form

$$x = A^i b + u, \qquad b \in [b^-, b^+], \qquad u \in N(A)$$

A^i satisfying $AA^iA = A$ and $(AA^i)^T = AA^i$ (since x minimizes the quantity $\|Ax - b\|$ in the least-squares sense yielding $A^TAx = A^Tb, \forall b$) still such a value of x does not in general satisfy the above consistency condition. Robers and Ben-Israël (1970) suggested instead that the interval programming problem be written in the form

$$\max c^Tx$$

$$\text{subject to } d^- \leqq Fx \leqq d^+$$
$$g^- \leqq hx \leqq g^+$$

where F is of full row rank, beginning with the simple case where h is a row vector, and ending with the general case where h is assumed to be different. Similar approaches can also be found in Charnes, Granot and Phillips (1977) and also Charnes, Granot and Granot (1977). Lata and Mittal (1976) extended Ben-Israël and Robers technique to the case when the objective function is given in the linear fractional form $(c^Tx + c_0)/(d^Tx + d_0)$. Here, we will limit ourselves to the above simple case, assuming that F is nonsingular as well in order to give the reader an idea about one of the approaches.

First, a transformation is sought in the form

$$Fx = z$$

whereby the interval programming problem is reduced to the following one

$$\max c^TF^{-1}z$$

$$\text{subject to } d^- \leqq z \leqq d^+$$
$$g^- \leqq hF^{-1}z \leqq g^+$$

z is therefore a feasible solution for the reduced problem, if the inequalities are satisfied. These inequalities, one should note, replace the condition for consistency mentioned before. To visualize this equivalence, write

$$A = \begin{bmatrix} F \\ \hline h \end{bmatrix}, \qquad b = \begin{bmatrix} d \\ \hline g \end{bmatrix}, \qquad A^i = [F^{-1} : 0]$$

Hence

$$AA^ib \subseteq [b^-, b^+], \qquad b \in [b^-, b^+]$$

implies that

$$hF^{-1}d \subseteq [g^-, g^+]$$
$$d \in [d^-, d^+]$$

or that

$$g^- \leqq hF^{-1}z \leqq g^+$$

where

$$d^- \leqq z \leqq d^+$$

Now the maximum value for $c^T F^{-1} z$ is obtained by observing as before the signs of $(c^T F^{-1})_i$. The optimal solution z^0 is therefore given by

$$z_i^0 = \begin{cases} d_i^+ , & (c^T F^{-1})_i > 0 \\ d_i^- , & (c^T F^{-1})_i < 0 \\ \theta_i d_i^+ + (1 - \theta_i)\, d_i^- , & (c^T F^{-1})_i = 0 , \quad 0 \leq \theta \leq 1 . \end{cases}$$

this provided of course that z^0 is feasible, i.e. $g^- \leqq hF^{-1}z \leqq g^+$. In case z^0 is not feasible, say

$$hF^{-1}z^0 > g^+$$

then choose $z^0 = z^*$, where z^* satisfies

$$hF^{-1}z^* = g^+$$

To do this efficiently, let

$$\Delta = g^+ - hF^{-1}z^0$$

be the amount by which the feasibility condition of z^0 is violated. It is clear that the only components of z_i^0 which have to be changed in order to ensure feasibility, are indexed by

$$i : 1 \leqq i \leqq r , \qquad (hF^{-1})_i \neq 0 , \qquad v_i = \frac{(c^T F^{-1})_i}{(hF^{-1})_i} \geqq 0$$

v_i is called the *marginal cost of moving towards feasibility*.

A simple example reproduced from Robers and Ben-Israël (1970) will now be used to visualize the above method:

$$\max x_1 + 2x_2$$

such that

$$-9 \leqq -3x_1 + x_2 \leqq 9$$
$$0 \leqq x_2 \leqq 8$$
$$2 \leqq x_1 + x_2 \leqq 6$$

We have

$$F = \begin{bmatrix} -3 & 1 \\ 0 & 1 \end{bmatrix}, \qquad d^+ = \begin{bmatrix} 9 \\ 8 \end{bmatrix}, \qquad d^- = \begin{bmatrix} -9 \\ 0 \end{bmatrix}$$

$$h = [1 \ \ 1] , \qquad g^+ = 6 , \qquad g^- = 2$$

The reduced problem is

$$\max \left[-\frac{1}{3}, \frac{7}{3}\right]\begin{bmatrix} z_1 \\ z_2 \end{bmatrix}$$

such that

$$\begin{bmatrix} -9 \\ 0 \end{bmatrix} \leq \begin{bmatrix} z_1 \\ z_2 \end{bmatrix} \leq \begin{bmatrix} 9 \\ 8 \end{bmatrix}$$

and

$$2 \leq \left[-\frac{1}{3}, \frac{4}{3}\right]\begin{bmatrix} z_1 \\ z_2 \end{bmatrix} \leq 6$$

The optimal solution is

$$\begin{aligned} z_1^0 &= -9 \\ z_2^0 &= 8 \end{aligned}$$

It is however infeasible, since

$$\left[-\frac{1}{3}, \frac{4}{3}\right]\begin{bmatrix} -9 \\ 8 \end{bmatrix} > 6$$

Calculate

$$\begin{aligned} v_1 &= (-1/3)/(-1/3) = 1 > 0 \\ v_2 &= (7/3)/(4/3) = 7/4 > 0\,, \end{aligned}$$

Hence both z_1^0 and z_2^0 have to be changed. Now

$$\Delta = 6 - \frac{41}{3} = -7\frac{2}{3},$$

and z_1^0 and z_2^0 are changed by amounts δ_1 and δ_2 such that

$$\delta_1\left(-\frac{1}{3}\right) + \delta_2\left(\frac{4}{3}\right) \leq \Delta = -7\frac{2}{3}$$

But since $v_1 < v_2$, a change in z_1^0 has first to be sought, by which $\delta_1 = 18$, yielding $\delta_2 = -\frac{5}{4}$. Hence

$$z_1^0 = -9 + 18 = 9$$

$$z_2^0 = 8 - \frac{5}{4} = \frac{27}{4}$$

and the optimal solution is therefore $x^0 = \left(-\frac{3}{4}, \frac{27}{4}\right)^T$.

Despite its being concerned specifically with the problem discussed in this section, interval programming may also be used in the treatment of linear programming problems where the cost coefficients are specified in interval form

$$c_i = [\underline{c}_i, \overline{c}_i], \qquad i = 1, \ldots, n$$

This class of problems falls in the category of *multiparametric programming*. They differ from the problems investigated in Sect. 5.3 in that multiple degrees of freedom are allowed when varying the objective function coefficients, or that in other words, as Steuer (1981) phrased it: it overcomes the *one at a time, hold everything else constant* situation. For problems of this type and their treatment, the reader may refer to Gal and Nedoma (1972) and also to Steuer (1976).

However, if the range of c_i falls within its decision region, the problem becomes an exercise in solving interval linear equations. In such case, one can consider the general case where A, b and c are all given in interval form. Much more generally still, one can incorporate the effect of round-offs, see in this respect Stewart (1973). Hence for the problem

$$\max \sum_{i=1}^{n} c_i^I x_i$$

subject to

$$x_i \geqq 0$$
$$A^I x = b^I ,$$

definition of the basis matrix B^I and its approximate mid-point inverse Y entails that (Moore (1979) p. 88)

$$x_i^I = x_i + r[-1, 1], \qquad x_i \text{ is a basic variable}$$
$$x_i^I = 0 \qquad\qquad\quad , \qquad x_i \text{ is a nonbasic variable}$$

where x is an approximate solution of $Bx = b$, for some $B \in B^I$ and $b \in b^I$ where also

$$r = \frac{\|Y\| \, \|B^I x - b^I\|}{1 - \|I - YB^I\|}$$

The bound for x^I can be further sharpened if $w(A^I)$ and $w(b^I)$ are small. One can call x^I an initial $z^{I(0)}$ and iterate, as in Sect. 3.7, using

$$z^{I(k+1)} = z^{I(k)} \cap \{Yb^I + (I - YB^I)\, y^{I(k)}\}, \qquad k = 1, 2, \ldots$$

yielding a nested sequence that contains the interval solution of the linear programming problem. For problems of solvability in interval linear programming, the reader is referred to Rohn (1981 b).

5.5 Programming Under Uncertainty

In many practical applications, the coefficients in a linear programming problem are random. For instance, price fluctuations of commodities assume a certain distribution around some mean or expected value. It is therefore natural to attribute to errors an underlying probabilistic distribution. Mandasky (1962) classified the solution methods of linear programming under uncertainty under three different titles: the *expected value* solution, the *fat* solution and the *slack* solution.

In the expected value solution, the stochastic linear programming in which the data are random is transformed into a non-stochastic problem by replacing the data by their respective expected values. As an example, consider in a diet problem that x_j is the quantity purchased of the jth food, p_j its price per unit of weight, a_{ij} the quantity of the ith nutrient contained in a unit of j, and b_i the minimum quantity required to maintain a good health. Then if the price of x_j obeys a distribution p_j with an expected price $\bar{p}_j$, the linear programming becomes

$$\min \sum_{j=1}^{n} \bar{p}_j x_j$$

subject to the conditions

$$x_j \geqq 0\,, \qquad j = 1, \ldots, n$$

$$\sum_{j=1}^{n} a_{ij} x_j \geq b_i\,, \qquad i = 1, \ldots, m$$

Hence, in the expected value solution, the expected prices are used in place of the distribution of prices, with the usual method of solution carried out. In the sequel, we shall denote the expected or mean value $E(x)$ for any distribution function $f(x)$ of a random element x. In other words, we have

$$E(x) = \int_{x^I} x f(x)\, dx$$

where the integration is carried out over the interval $x^I = [\underline{x}, \overline{x}]$ which is the range of variation of x. x^I provides the domain of uncertainty of x and, according to Dempster (1969), a support for x, or *supp* x.

In fact, the above mean or center form of x is also suitable for solving the linear equations $Ax = b$, when A and b are random. Hansen's method for handling interval equations (cf. Sect. 2.3 and Sect. 3.7) can be formulated as

$$x^{(k+1)} = \{I - BE(A\ \ )\}\, E(x^k) + BE(b)$$

where B is an approximate inverse of A.

However, the expected value solution has its drawbacks. For the above linear programming problem, if $\mathring{F}(A, b, c)$ is the optimal value of F, then

$$E\{\mathring{F}(A, b, c)\} \neq \mathring{F}\{E(A), E(b), E(c)\}$$

in general. Hence, although the expected value solution provides a useful reference point, especially when the data scatter is small, it may as well lead to erroneous results in instances where the uncertainty is significant. Hanson (1960) provides a simple example, with only A random, wherein due to constraints conditioning a small variation in the parameters leads to an underestimation of $E(\mathring{F})$ by $\mathring{F}$, by more than 30%.

Another drawback triggered by large uncertainties in the data is that $E(\mathring{F})$ may not be finite. To visualize this, recall that each optimal solution x^k corresponds to a range of parameter values contained in the so-called characteristic region (see Sect. 5.3) or decision region, according to Bereanu (1967). For parameter variation within the kth decision region Δ_k, x^k remains optimal. The interiors of the decision regions are disjoint, and their union, $\Delta = \bigcup_k \Delta_k$, is a convex closed set of R^n. Hence whilst each parameter α ranges over its support α^I, the optimal solution x^k changes from one vertex to the other as we pass through many decision regions. We can therefore conclude that, for $E(\mathring{F})$ to be finite, the probability that α is contained entirely in all decision regions must be equal to unity, i.e.

$$Pr\{\text{supp}\,\alpha \in \Delta\} = 1$$

This is nonetheless not all what there is to it. If x^* minimizes $E(c^T)\,x$ subject to $x \geqq 0$ and $E(A)x \geqq E(b)$, it does not necessarily satisfy the condition $Pr(Ax^* \geqq b) = 1$. x^* is then termed *not-permanently feasible*. Furthermore, although x^* may be feasible, the dual programming problem may not be optimized by y^*. Rather, both vectors x^* and y^* are feasible (and hence optimal) expected value solutions of the primal and dual problems if the first m elements of the vector Qz are nonnegative (see Mandasky (1962)). z is the vector of the optimal strategy for the matrix game $E(Q)$, where Q is the pay-off matrix

$$\begin{bmatrix} 0 & A & -b \\ -A^T & 0 & c \\ b^T & -c^T & 0 \end{bmatrix}$$

For a proof of the equivalence between game theory and linear programming, the reader may refer to an early paper by Dantzig (1951).

However, if the uncertainties in the data are small, the problem becomes much easier. A first approach to the general distribution problem is to assume that the support of the random vector formed from A, b and c lies in one single decision region. Such an assumption is convenient for carrying out a sensitivity analysis for small errors. Such an approach was first adopted by Babbar (1955) who expressed the optimal random vector x^0 in terms of random determinants using Cramer's rule, then approximating the distribution of these determinants by normal distributions (see Sect. 5.6). A more systematic and direct approach was developed by Prékopa (1966) who expanded $\mathring{F}(A, b, c)$ around $E(\mathring{F})$ using Taylor's series when $E(A)$, $E(b)$ and $E(c)$ belong to a decision region. It then follows from Sect. 5.2 that

$$F^0 - E(F^0) \approx -y^*(B - E(B))\,x^* + (c - E(c))^T\,x^* + (b - E(b))^T\,y^*$$

for small random distribution around the expected value solution.

In the fat solution method, the random elements are replaced by pessimistic estimates of their values. This method is usually used when the expected value solution has a high degree of infeasibility. Hence, by postulating a pessimistic (A, b) and solving the non-stochastic problem, we may obtain the degree of feasibility required. A vital question: how to determine appropriate pessimistic values for the random elements, so that x^0 will be optimal and permanently feasible?

We first note that for each (A, b), the values of x that satisfy $Ax \geqq b$, $x \geqq 0$ form a convex polyhedron. The set of permanently feasible x's is formed by the intersection of these polyhedra; it is also a convex set. Hence, by indexing the possible values of (A, b) as $(A^{(r)}, b^{(r)})$; $r = 1, \ldots, R$, the permanently feasible values of x satisfy $x \geqq 0$, $A^{(r)}x \geqq b^{(r)}$, $r = 1, \ldots, R$. The solution of the stochastic programme is then an x that minimizes $E(c)^T\ x$ subject to the conditions $x \geqq 0$ and $A^{(r)}x \geqq b^{(r)}$, $r = 1, \ldots, R$. Instead, if (A, b) has a continuous distribution it can be transformed into the above discrete one by sampling over its support (A^I, b^I). However, if only b is random, then $Ax \geqq \bar{b}$, $x \geqq 0$ define a subset of the set of permanently feasible x's. Here $\bar{b}$ is the vector of which the ith coordinate is the supremum of the ith coordinate of the possible vector b.

Still, both the fat and expected value solutions suffer from serious drawbacks. In the fat solution, the random variations in the elements are ignored, while we provide plenty of *fat* in the deterministic version of the problem, with the hope that the uncertainty will be completely absorbed. But this usually leads to results that are far from being optimal. In the expected value solution, although results do not suffer from such inaccuracies, the solution obtained may be far from being feasible. The slack solution, on the contrary, is always feasible, and handles the inherent uncertainty at the same time.

The slack solution, also called the *two-stage solution* originated in the context of planning production according to a varying demand. The term "two-stage solution" is due to its involving a decision x to be made first, after which the random elements are observed and a second decision y is made. Note that the quantities of activities x in the first stage are the only ones to be determined, while those of the second stage y are determined later. According to Dantzig (1955), the set of activities is called *complete*, in the sense that whatever choice of activities x is made, there is always a possibility of choosing y. Hence, it is not possible, as Dantzig concluded, to get in a position where the programming problem admits no solution. To illustrate what a slack solution means, we recall the author's very example: a factory has 100 items on hand which may be shipped to an outlet at the cost of \$ 1.00 a piece to meet an uncertain demand d. If the supply exceeds the demand, the unused stock will have no value. But suppose the demand is larger, then in order to meet the unsatisfied demand, the price could be raised locally at \$ 2.00 a piece. In this case, the equations would look like:

$$100 = x_1 + x_2$$
$$d = x_1 + y_1 - y_2$$
$$F = x_1 + 2y_1$$

where

x_1 = number of pieces shipped
x_2 = number of pieces in stock
d = demand (yet unknown)
y_1 = number of pieces purchased on the open market
y_2 = variable denoting the excess of supply over demand

Note that $x_2 + y_2$ can be considered of no value or written-off at some very reduced price at a later stage. In the automobile market, for instance, the past year's models are sold rather cheap. In the coffee market, excess is always burnt. Note that this problem always admits a solution under an uncertain demand d. The reason is that y_1 and y_2 have opposite signs, whence the slacks are free to vary with a varying demand. After x_1 is determined from the first stage, y_1 and y_2 are calculated, i.e. allocations in the first stage are made to meet an uncertain demand occurring in the second stage.

In the above simple problem, we had only one product. But, in general, we have to assign various resources x_{ij} to several destinations j, i being the number of different resources. Let also b_{ij} represent the number of units of demand at destination j that can be satisfied by one unit of resource i. If u_j represents the total amount of the resources shipped at destination j, v_j the shortage of supply and s_j the excess of supply, then
First stage:

$$\sum_{j=1}^{n} x_{ij} = a_i$$

$$\sum_{i=1}^{m} b_{ij} x_{ij} = u_j$$

Second stage:

$$d_j = u_j + v_j - s_j\,, \qquad j = 1, \dots, n$$

Total cost:

$$F = \sum_{i=1}^{m} \sum_{j=1}^{n} c_{ij} x_{ij} + \sum_{j=1}^{n} \alpha_j v_j$$

The problem here is that of minimizing F which is composed of two parts: the costs of assigning the resources to the destinations as well as the costs (i.e. lost revenues) incurred because of the failure of the total amounts u_j, $j = 1, \dots, n$ to meet the unknown demands d_j, $j = 1, \dots, n$. Note that F depends linearly on x_{ij} and v_j, which in turn depend on u_j and d_j. The problem is usually set so as to minimize $E(F)$. The latter is a convex function in u_j (cf. Dantzig (1955)). Unfortunately, it may also be nonlinear, thus impairing the use of linear programming. But in many practical problems the objective function can be represented by the sum of separable convex functions, case in which a linear approximation can be carried out (cf. Charnes and

Lemke (1954) and Dantzig (1955)). In case the demand function is of discrete distribution, no approximation is required (cf. Elmaghraby (1959)). For further reading on the subject and related topics, the reader may refer to Wets (1964, 1965a, 1965b).

The two-stage problem in linear programming can be formulated in the following concise form due to Mandasky (1962), namely:

$$\text{Min } E(c_1^T x + c_2^T y)$$

$$\text{subject to } Ax + By = b$$
$$x, y \geqq 0$$

where A is a random $m \times n_1$ matrix of known distribution, B a known $m \times n_2$ matrix, b a random m-dimensional vector of known distribution and c_1 and c_2 are known cost coefficients of dimension n_1 and n_2 respectively. Note that By stands for $y^+ - y^-$ as before, y^+ representing the excess demand with respect to supply and y^- the excess of supply over the demand. In other words, $y^+ = b - Ax$ and $y^- = 0$ if $b \geqq Ax$, whereas $y^- = Ax - b$ with $y^+ = 0$ if $Ax > b$. This problem has always a feasible solution, since y^+ and y^- are chosen freely from the algorithm. This is why we call x a *decision* vector and y a *slack* vector. Then the assumption that $Ax + By = b$ is satisfied by (x, y) irrespective of (A, b) is equivalent to saying that after the decision has been made and the subsequent random event has been observed, one can always compensate with a slack y depending on x, A, B and b for inaccuracies in the decision.

The two-stage problem in linear programming was introduced by Dantzig (1955) as a method for allocating different resources to an uncertain demand. The theory has since been proving useful in a variety of applications, starting from the classical problem of aircraft-routing allocation studied by Ferguson and Dantzig (1956) and ending with the planning of finance under uncertainty (see Kallberg et al. (1982)).

5.6 Distributions of Solution

Except in very special cases, the solution of the system of linear random equations $Ax = b$, in which each of $\{a_{ij}, b_i, i = 1, \dots, m, j = 1, \dots, n\}$ is a random variable with a given distribution, is difficult to obtain or describe. True, in regression analysis, A and b can be both random (usually b only is random), but x is assumed non-stochastic and drawn from the estimator $\hat{x}$ using confidence intervals pertaining to the sampling distribution. The philosophy behind regression methods is to achieve an average relationship for the population under study assuming already a form of this relation (be it a straight-line for the case of linear regression) where one seeks to determine this mean or average. The departure therefore from this relation is presupposed to be caused by errors in the measurements which have to be smoothed out using regression. Contrary to this point of view is the case in which this departure is considered to be due to the genuine variability in the material itself like in most physical and biological models. The reason is that the parameters involved in the mathematical equations describing the physical model are known only as random variables. It would be interesting therefore to obtain the distribution of the solution to study further on their statistical properties. For the simple linear equations $Ax = b$ discussed hereunder, the problem is to obtain-if possible-the distribution of x given the distribution of the elements of A and b.

Although the problem looks simple, it is by no means trivial. Even if all random variables are independent, each component x_j is a ratio involving sums and products (from Cramer's rule). To find the density function of x_j, we need to determine therefore the density of each product, the density of each sum (the terms of the sum are not independent) and then the density of the ratios (again not independent). In principle, all this is straightforward but computationally it is not to the least simple. Approximate solutions can, however be found if the variances of all variables are small. Another attempt is to seek not densities but only moments.

Among the early attempts to tackle the above problem was the one of Babbar (1955). The latter investigated the distribution of solutions to the linear equations

$$(A + \delta A)\, x = b + \delta b$$

where A and b are a constant matrix and *a* constant vector and are both known. δA and δb are respectively known matrix and vector of random errors having

$$E(\delta a_{ij}) = 0\,, \qquad E(\delta^2 a_{ij}) = \sigma_{ij}^2$$

and

$$E(\delta b_i) = 0\,, \qquad E(\delta^2 b_i) = \sigma_i^2$$

The problem therefore is to obtain approximate distributions for the variables $x_1, \ldots, x_n$. Babbar extended his analysis to the linear programming problem in which the vectors of A become the basis of the solution. Hence the objective function will reach an optimum of $F = c^T x_{\text{opt}}$, where $c^T = (c_1, \ldots, c_n)$ is the cost coefficient vector. The latter can also be taken to be random with $E(\delta c_i) = 0$ and $E(\delta^2 c_i) = \omega_i^2$.

Now, if $A + \delta A$ is nonsingular, in other words $Pr[\det (A + \delta A) \neq 0] = 1$, we have that

$$x_k = \frac{\det (A^k + \delta A^k)}{\det (A + \delta A)}\,, \qquad F = \sum_{k=1}^{n} (c_k + \delta c_k)\, x_k$$

where $A^k + \delta A^k$ is equal to $A + \delta A$ except that the kth column is substituted for by $b + \delta b$. But we have to a first order approximation that

$$\det (A^k + \delta A^k) \cong \det (A^k) + \sum_{i=1}^{n} \delta b_i C_{ik} + \sum_{i=1}^{n} \sum_{\substack{j=1 \\ j \neq k}}^{n} \delta a_{ij} C_{ij}^k = N(x_k) \text{ say}$$

and

$$\det (A + \delta A) \cong \det (A) + \sum_{i=1}^{n} \sum_{j=1}^{n} \delta a_{ij} C_{ij} = D(x_k) \text{ say}$$

where C_{ij} is the cofactor of the element a_{ij} and C_{ij}^k is the cofactor of a_{ij} in the matrix A^k (A^k is equal to A except that its kth column is replaced by the vector b). It follows that the mean, variance and covariance of the above functions are

$$E[\det(A^k + \delta A^k)] \cong \det(A^k)$$

$$V[\det(A^k + A^k)] \cong \sum_{i=1}^{n} \sigma_i^2 C_{ik}^2 + \sum_{i=1}^{n} \sum_{\substack{j=1 \\ j \neq k}}^{n} \sigma_{ij}^2 (C_{ij}^k)^2$$

$$E[\det(A + \delta A)] \cong \det(A)$$

$$V[\det(A + \delta A)] \cong \sum_{i=1}^{n} \sum_{j=1}^{n} \sigma_{ij}^2 C_{ij}^2$$

and

$$\operatorname{Cov}[\det(A^k + \delta A^k), \det(A + \delta A)] \cong \sum_{i=1}^{n} \sum_{\substack{j=1 \\ j \neq k}}^{n} \sigma_{ij}^2 C_{ij} C_{ij}^k$$

The reader interested in means of random determinants may also consult the short note by Bellman (1955).

Our task focusses next on the distribution of x_k knowing those of $N(x_k)$ and $D(x_k)$. This can be accomplished using the joint distribution of the errors involved or by employing characteristic functions and inversion formulas. However if the errors are normally distributed, the problem becomes much easier. Babbar made use of a theorem provided by Geary in 1930 relating the distribution of a quotient of two normal variates to their means, variances and covariances. Geary showed that if N and D are normally distributed variables having respectively means, variances and covariance $E(N)$, $E(D)$, σ_N^2, σ_D^2 and σ_{ND}, then the variable

$$t = \frac{E(D)\, Z - E(N)}{(\sigma_N^2 - 2\sigma_{ND} Z + \sigma_D^2 Z^2)^{1/2}}$$

is normally distributed with mean zero and variance unity, where $Z = N/D$, provided that $E(D) > 3\sigma_D$. Babbar used this theorem to show that the probability distribution of the quotient Z comes as

$$f(Z)\, dZ = \frac{1}{\sqrt{2\Pi}} \frac{(E(D)\, \sigma_N^2 - E(N)\, \sigma_{ND}) + Z(E(N)\, \sigma_D^2 - E(D)\, \sigma_{ND})}{(\sigma_N^2 - 2\sigma_{ND} Z + \sigma_D^2 Z^2)^{3/2}}$$
$$\times \exp\left(-\frac{1}{2}(ZE(D) - E(N))^2/\sigma_N^2 - 2\sigma_{ND} Z + \sigma_D^2 Z^2\right) dZ$$

And by substituting respectively for $E(N)$, $E(D)$, σ_N^2, σ_D^2 and σ_{ND} by $E[\det(A^k + \delta A^k)]$, $E[\det(A + \delta A)]$ etc., we obtain the distribution of x_k. Similar arguments can be brought up for evaluating the corresponding distribution of the optimal objective function F associated with the linear programming problem. For this, the reader is referred to Babbar's paper. The idea of approximating solutions of equations using perturbation bounds when the variables are uncorrelated have been a great deal implemented in engineering, see for instance Papoulis (1966).

Furthermore, Babbar investigated probability limits of the solution x_k in other words of the quotient $Z = N/D$, from which he obtained two limits of Z to lie between with a certain prescribed confidence level. The same idea of confidence limits can also be found in Kuperman (1971, Chap. 12). The latter further suggested a *statistical ill-conditioned factor* of the equations when being less than one-subject to some confidence limits-will ensure that the random variations in A will not bring det (A) to zero. In other words, a system of linear equations are said critically ill-conditioned if the determinant of the coefficient matrix can become zero within the limits of the uncertainties in the coefficients (cf. Sect. 1.3).

Since the early work of Babbar, the theory of random equations has attracted a great deal of interest among workers. A reasonably unified treatment regarding the operator equation

$$A(\omega)\, x = b(\omega)$$

where ω is a random variable is provided by Bharucha-Reid (1964, 1979). Existence, uniqueness and measurability of the solution x are established and also some methods have been outlined. For example in Nashed and Engl (1979) the authors proposed a method of successive approximations like the steepest-descent method to minimize the quadratic index

$$J(\omega, x) = \frac{1}{2}\, \|A(\omega)\, x - b(\omega)\|_2^2$$

especially suited to solve the least-squares problem when A has dimension $m \times n$ $(m > n)$. By defining a sequence of random approximants $x^0(\omega)$, $x^1(\omega)$, ... according to the relation

$$x^{n+1}(\omega) = x^n(\omega) - \alpha_n \quad \text{grad. of} \quad J(\omega, x^n(\omega))$$

in which α_n is chosen to minimize $J(\omega, x^{n+1}(\omega))$, i.e.

$$\alpha_n(\omega) = \frac{\|r^n(\omega)\|^2}{\|A(\omega)\, r^n(\omega)\|^2}$$

where $r^n(\omega) = A^*(\omega)\, A(\omega)\, x(\omega) - A^*(\omega)\, b(\omega)$, $x^n(\omega)$ converges to the random solution of $A(\omega)\, x = b(\omega)$ given by $x(\omega) = A^+(\omega)\, b(\omega)$. The authors further showed that the rate of convergence of the iterative scheme is geometric with ratio $(k^2(\omega) - 1)/(k^2(\omega) + 1)$ where $k(\omega) = \|A(\omega)\|\, \|A^+(\omega)\|$ is the pseudo-condition number of the operator $A(\omega)$. For an outline of the method of steepest-descent as applied to solve singular linear operator equations, the reader is to consult Nashed (1970). And for a reading about gradient methods for solving linear operator equations, the paper of McCormick (1975) is recommended.

Nake (1967) on the contrary followed a direct scheme to solve the linear system $A(\omega)\, x + b(\omega) = 0$ using a joint distribution function of the random variables

$b_1(\omega), \dots, b_n(\omega), a_{11}(\omega), \dots, a_{nn}(\omega)$. If $\Phi_{n, n^2}(z, Z) = \Phi_{n, n^2}(z_1, \dots, z_n, z_{11}, \dots, z_{nn})$ is such function, then the distribution of the solution is given by the n-dimensional distribution function

$$\mu_n(\xi) = \int |\det (Z)| \cdot \Phi_{n, n^2}(-Z.\xi, Z)\, dZ .$$

Though difficult generally to evaluate, the above integral can be dealt with under special cases like in Nake's thesis, when the variables are uncorrelated, $A(\omega)$ is a triangular matrix and also assumes each a normal distribution.

The above discussed methods whether direct or iterative remain far impractical as the reader must have already noticed. A very practical method to handle linear random operator equations is the experimental Monte Carlo method. If the elements of $A(\omega)$ and $b(\omega)$ are independent and assume each some distribution function over an interval, then by generating a sequence of random numbers inside these intervals to simulate the available distribution and solving each time for the vector x, one obtains after some large number of runs the set X of approximately all feasible solutions. It will be found that $x(\omega)$ will assume a normal form whenever a_{ij} and b_i independently do. The reader interested in Monte Carlo methods, in dealing with linear operator equations, is referred to Hammersley and Handscomb (1964) and also Raj (1980).

Summarizing the subject of sensitivity analysis in linear programming, as it was discussed in the foregoing chapters, we quote Rappaport's conclusion (Rappaport (1967)). Sensitivity analysis has the following advantages:

1. It helps to determine the responsiveness of the conclusions of an analysis to changes or errors in parameter values used in the model.
2. It tests the responsiveness of model results to passive changes in parameter values, and thereby offers valuable information for appraising the relative risk among alternative courses of action.
3. It provides systematic guidelines for allocating scarce organizational resources to data collection and data refinement activities.
4. If the value of a decision is insensitive to estimated parameter variations, this determines a decision region in which the decision is valid.
5. If the value of a decision is sensitive to estimated parameter variations, a statistical model can be developed to guide the information decision.

Exercises 5

1. Show that

$$(A + uv^T)^{-1} = A^{-1} - \frac{A^{-1}uv^TA^{-1}}{1 + v^TA^{-1}u}$$

for any two vectors u and v. Hence, use the result to show that

$$(B + \Delta a_{ij}E)^{-1} = B^{-1} - \frac{\Delta a_{ij}}{1 + \Delta a_{ij}a'_{ji}} B^{-1}EB^{-1}$$

where E is a matrix with $e_{ij} = 1$ and zero otherwise; a'_{ji} is the element (j, i) in the final tableau.

2. If in the problem

$$\min F = -20x_1 - 10x_2 - 20x_3$$

subject to $x_1, x_2, x_3 \geqq 0$

$$3x_1 + x_2 + 4x_3 \leqq 480$$
$$4x_1 + 2x_2 + 3x_3 \leqq 400$$
$$x_2 \leqq 70$$

the optimal solution is $x^* = (112, 32, 38)$, find the range of the cost coefficients c_1, c_2, c_3 so that x^* will remain optimal. Then obtain the range of $F = [\underline{F}, \overline{F}]$ within the decision regions of the cost coefficients as an interval function. Calculate also $\partial F/\partial c_1$, $\partial F/\partial b_3$ and $\partial F/\partial a_{11}$, where a_{11} is the first entry in the base matrix. Obtain $F(c_1 + \varepsilon c_1)$ where ε is sufficiently small.

3. m measurements are taken to determine the variables $x_1, \dots, x_n$, where $m > n$, subject to $Ax = b$. A famous technique is to calculate the value of x which minimizes $\|Ax - b\|$ under some norm (least-squares fitting). Show that $Ax \leqq b$ defines a convex set C in $\mathbb{R}^n$ and obtain the range of x^I which covers minimally the set C. Hint: use linear programming and maximize or minimize x, so that $Ax \leqq b$. Show that $Ax^I \supseteq b$.

4. If in the problem $\min F = x_1 - 3x_2 + 2x_3$, subject to

$$x_1, x_2, x_3 \geqq 0$$
$$3x_1 - x_2 + 2x_3 \leqq 7$$
$$2x_1 - 4x_2 \leqq 12$$
$$-4x_1 + 3x_2 + 8x_3 \leqq 10$$

each coefficient c_1, c_2, c_3 undergoes a perturbation of $+10\%$, each coefficient of A a perturbation of -20% and each of those of b a perturbation of $+30\%$, calculate the new optimal solution and the final objective function. Perform a sensitivity analysis with respect to the cost coefficients.

5. Solve the interval problem

$$\max x_2 - x_1$$

subject to the conditions

$$0 \leqq x_1 \leqq 4$$
$$0 \leqq x_2 \leqq 6$$
$$3 \geqq x_1 + x_2 \geqq 1 .$$

6. Solve the interval problem

$$\max c_1x_1 + c_2x_2$$

subject to the relations

$$1 \leqq x_1 + x_2 \leqq 3$$
$$2 \leqq x_3 \leqq 4\,.$$

Show that $x = A^i b + u$, where $u \in N(A)$. Hence, obtain a relation between c_1 and c_2 such that F^0 be finite; calculate F^0.

7. Solve the interval problem

$$\max x_1 + 2x_2$$

if $0 \leqq x_1 \leqq 6$
$0 \leqq x_2 \leqq 8$
$-9 \leqq -3x_1 + x_2 \leqq 9$
$2 \leqq x_1 + x_2 \leqq 6$

8. For the problem

$$\min\,(30 + 6t)\,x_1 + (50 + 7t)\,x_2$$

subject to the conditions

$$(14 + 2t)\,x_1 + (4 + t)\,x_2 \leqq 14 + t$$
$$(150 + 3t)\,x_1 + (200 + 4t)\,x_2 \geqq 200 + 2t$$
$$x_1, x_2 \geqq 0$$

obtain

$$\frac{dx_1}{dt}\,, \quad \frac{dx_2}{dt} \quad \text{and} \quad \frac{dF}{dt} \quad \text{at all } t \text{ governed by } t \geqq 10\,.$$

9. Draw a similarity between the two problems
a) Min $c^T x$, with $Ax \geqq b$, $x \geqq 0$
Max $b^T y$, with $A^T y \leqq c$, $y \geqq 0$
b) $Ax = b$, with $A^T y = e$, $e = (0, 0, \dots, 0, 1_k, 0, \dots, 0)$

Show that

$$\frac{\partial x_k}{\partial a_{ij}} = -y_i^0 x_j^0 / c_k$$

for the first problem is equivalent to

$$\frac{\partial x_k}{\partial a_{ij}} = -y_i x_j \quad \text{for the second problem.}$$

10. If for the problem

$$\min 30x_1 + 50x_2$$

subject to the conditions

$$x_1, x_2 \geqq 0$$
$$14x_1 + 4x_2 \leqq 14$$
$$150x_1 + 200x_2 \geqq 200$$

The term a_{22} becomes $200 + \lambda$, find the characteristic region of λ for x_1, x_2 to remain optimal.

11. In the above problem, a_{11} becomes $14 + \mu$, obtain the characteristic region in μ and λ. And if λ is restricted to the interval [0, 7], then define the interval values of μ which correspond to as many as possible of the extreme points x^0.

12. Solve the linear interval programming problem

$$\min [1 \pm 0.1]\, x_1 - [3 \pm 0.2]\, x_2 + [2 \pm 0.1]\, x_3$$

such that

$$x_1, x_2, x_3 \geqq 0$$
$$3x_1 - x_2 + 2x_3 \leqq [7 \pm 0.1]$$
$$2x_1 - [4 \pm 0.1]\, x_2 \leqq 12$$
$$-4x_1 + 3x_2 + [8 \pm 0.2]\, x_3 \leqq 10\,.$$

References

Albert A (1972) Regression and the Moore-Penrose pseudo-inverse. Academic Press, New York

Albert A (1976) Statistical applications of the pseudo-inverse. In: Nashed M (ed) Generalized inverses and applications. Academic Press, New York

Alefeld G, Herzberger J (1983) Introduction to interval computations. Academic Press, New York

Alefeld G (1984) On the convergence of the higher order versions of D J Evans' implicit matrix inversion process. ZAMM 64: 413–418

Anderson N, Karasalo I (1975) On computing bounds for the least singular value of a triangular matrix. BIT 15: 1–4

Anderson D, Sweeney D, Williams T (1976) An introduction to management science. West. Publ. Minnesota

Anderson W, Trapp G (1985) Inverse problems for means of matrices, 2nd Siam Conf. on applied linear algebra, Raleigh, NC. To appear also in: Siam J Alg & Discrete methods

Atiqullah M (1969) On a restricted least-squares estimator. J Amer Statist Assoc 64: 964–968

Babbar M (1955) Distributions of solutions of a set of linear equations (with an application to linear programming). J Amer Statist Assoc 50: 854–869

Bartels R, Stewart G (1972) A solution of the equation $AX + XB = C$. Commun ACM 15: 820–826

Barth W, Nuding E (1974) Optimale Lösung von Intervallgleichungssystemen. Computing 12: 117–125

Bauer F (1963) Optimally scaled matrices. Numer Math 5: 78–87

Bauer F, Heinhold J, Samelson K, Sauer R (1965) Moderne Rechenanlagen, Chap. 3. Teubner, Stuttgart

Bauer F (1966) Genauigkeitsfragen bei der Lösung linearer Gleichungssysteme. ZAMM 46: 409–421

Beckenbach E, Bellman R (1965) Inequalities. Springer, New York

Beeck H (1974) Zur Scharfen Außenabschätzung der Lösungsmenge bei Linearen Intervallgleichungssystemen. ZAMM 54: T208–209

Beeck H (1975) Zur Problematik der Hüllenbestimmung von Intervallgleichungssystemen. In: Nickel K (ed) Interval mathematics. Springer, Berlin, Heidelberg, New York (Lecture Notes in Computer Science Vol. 29)

Bellman R (1955) A note on the mean value of random determinants. Quart Appl. Math 13: 332–324

Bellman R (1970) Introduction to matrix analysis. McGraw-Hill, New York

Ben-Israël A (1966) On error bounds for generalized inverses. Siam J Numer Anal 3: 585–592

Ben-Israël A, Charnes A (1968) An explicit solution of a special class of linear programming problems. Opns Res 16: 1166–1175

Ben-Israël A, Robers P (1970) A decomposition method for interval linear programming. Management Sc 16: 374–387

Ben-Israël A, Greville T (1974) Generalized inverses: Theory and applications. John-Wiley, New York

Bereanu B (1967) On stochastic linear programming distribution problems. Stochastic technology matrix. Z Wahrscheinlichkeitstheorie verw Geb 8: 148–152

Berkson J (1950) Are there two regressions. J Amer Statist Assoc 45: 164–180

Bharucha-Reid A (1964) On the theory of random equations. In: Bellman R. (ed) Stochastic processes in mathematical physics and engineering. Proc Symp Appl Math, XVI, Amer Math Soc

Bharucha-Reid A (1979) Approximate solution of random equations. North-Holland, New York

Bierbaum F (1974, 1975): Intervall-Mathematik. Eine Literatur-Übersicht, Interner Berichte 74/2 and 75/3. Institut für Praktische Mathematik, Universität Karlsruhe

Björck A (1967) Iterative refinement of linear least-squares solutions I. BIT 7: 257–278

Björck A (1968) Iterative refinement of linear least-squares solutions II. BIT 8: 8–30

Björck A (1978) Comment on the iterative refinement of least-squares solutions. J Amer Statist Assoc 73: 161–166

Boot J (1963) On sensitivity analysis in convex quadratic programming problems. Opns Res 11: 771—786

Bouillon T, Odell P (1971) Generalized inverse matrices. John Wiley, New York

Branham R (1980) Least-squares solution of ill-conditioned systems II. Astron J 85: 1520–1527

Broyden C (1973) Some condition number bounds for the Gaussian elimination process. J Inst Math Appl 12: 273–286

Bunch J (1971) Equilibration of symmetric matrices in the max-norm. J Assoc Comp Mach 18: 566–572

Businger P, Golub G (1965) Linear least squares solutions by Householder transformations. Numer Math 7: 269–276

Calahan D (1972) Computer-aided network design. McGraw-Hill, New York

Campbell S, Meyer C (1979) Generalized inverses of linear transformations. Pitman, London

Carnahan B, Luther H, Wilkes J (1969) Applied numerical analysis. John-Wiley, New York

Casella G, Strawderman W (1980) Confidence bands for linear regression with restricted predictor variables. J Amer Statist Assoc 75: 862–868

Chan N (1982) Linear structural relationships with unknown error variances. Biometrika 69: 277–279

Chan L, Mak T (1984) Maximum likelihood estimation in multivariate structural relationships. Scand J Statist 11: 45–50

Charnes A, Lemke C (1954) Minimization of nonlinear separable functions, Graduate school of industrial administration, Carnegie Institute of Technology, see also Naval Res Log Quart 1: 301–312

Charnes A, Granot F, Phillips F (1977): An algorithm for solving interval linear programming problems. Opns Res 25: 688–695

Charnes A, Granot D, Granot F (1977) A primal algorithm for interval linear-programming problems. Linear Alg and its Appl 17: 65–78

Chartres B, Geuder J (1967) Computable error bounds for direct solution of linear equations. J Assoc Comp Mach 14: 63–71

Chew V (1970) Covariance matrix estimation in linear models. J Amer Statist Assoc 65: 173–181

Chipman J (1976) Estimation and aggregation in econometrics: An application of the theory of generalized inverses. In: Nashed M (ed) Generalized inverses and applications. Academic Press, New York

Cline A, Moler C, Stewart G, Wilkinson J (1979) An estimate for the condition number of a matrix. Siam J Numer Anal 16: 368–375

Cline A, Conn A, Van Loan C (1982) Generalizing the LINPACK estimator, In: Hennart J (ed) Numerical analysis Springer, Berlin, Heidelberg, New York (Lecture Notes in Mathematics Vol. 909)

Cochran W (1968) Errors of measurement in statistics. Technometrics 10: 637–666

Cochran W (1972) Some effects of errors of measurement on linear regression. Proc. Six*th* Berkeley Symp. on Math. Statist. and Prob., Univ. of Calif. Press, Vol. I

Cope J, Rust B (1979) Bounds on solutions of linear systems with inaccurate data. Siam J Numer Anal 16: 950–963

Courtillot M (1962) On varying all the parameters in a linear programming problem and sequential solution of a linear programming problem. Opns Res 10: 471–475

Cragg J (1966) On the sensitivity of simultaneous equations estimators to the stochastic assumptions of the models. J Amer Statist Assoc 61: 136–151

Curtis A, Reid J (1972) On the automatic scaling of matrices for Gaussian elimination, J Inst Math and its Appl 10: 118–124

Dahl G (1978) On scaling in linear algebraic systems. BIT 18: 363–365

Daniel J (1973) On perturbations in systems of linear inequalities. Siam J Numer Anal 10: 299–307

Dantzig G (1951) A proof of the equivalence of the programming problem and the game problem, In: Koopmans T (ed) Activity analysis of production and allocation. Wiley, New York

Dantzig G (1955) Linear programming under uncertainty. Management Sc 1: 197–206

Dantzig G (1963) Linear programming and extensions. Princeton Univ. Press, New York

Deif A (1981) Sensitivity analysis from the state equations by perturbation techniques. J Appl Math Model 5: 405–408

Deif A (1982) Advanced matrix theory for scientists and engineers. Halsted Press div., Wiley, New York

Deif A (1983a) Error bound for the equation $AX + XB = C$, IEEE Int. Symp. on circuits and systems (ISCAS), ed: Mitra, S., CA

Deif A (1983b) The generalized inverse of a perturbed singular matrix. ZAMP 34: 291–300

Demidovich B, Maron I (1973) Computational mathematics. Mir Publ., Moscow

Demko S (1985) Condition number of rectangular systems and bounds for generalized inverses, 2nd Siam Conf. on appl. Linear Alg., Raleigh. NC. To appear also in Linear Alg. and its Appl.

Demko S (1985) Spectral bounds for $\|A^{-1}\|_{\infty}$, to Appear in J Approx theory

Dempster M (1969) Distributions in intervals and linear programming In: Hansen E (ed) Topics in interval analysis. Clarendon Press, Oxford

Dempster A, Rubin D, Tsutakawa R (1981) Estimation in covariance components models. J Amer Statist Assoc 76: 341–353

Dent W, Hildreth C (1977) Maximum Likelihood estimation in random coefficient models. J Amer Statist Assoc 72: 69–72

Deutsch R (1965) Estimation theory. Prentice-Hall, N. J.

Dinkelbach W (1969) Sensitivitätsanalysen und Parametrische Programmierung. Springer, Berlin, Heidelberg, New York (Ökonometrie und Unternehmensforschung Bd. 12)

Director S, Rohrer R (1969a) The generalized adjoint network and network sensitivity. IEEE Trans Circuit Theory CT-16: 318–323

Director S, Rohrer R (1969b) Automated network design-The frequency domain case. IEEE Trans Circuit Theory CT-16: 330–337

Dixon J (1983) Estimating extremal eigenvalues and condition numbers of matrices. Siam J Numer Anal 20: 812–14

Dongarra J., Bunch J, Moler C, Stewart G (1979) LINPACK Users' Guide. Siam, Philadelphia

Duncan A (1974) Quality control and industrial statistics, 4th ed., Richard Irwin, Illinois

Dunn O (1968) A note on confidence bands for a regression line over a finite range. J Amer Statist Assoc 63: 1028–1033

Elden L (1980) Perturbation theory for the least-squares problem with linear equality constraints. Siam J Numer Anal 17: 338–350

El-Maghraby S (1959) An approach to linear programming under uncertainty. Opns Res 7: 208–216

Faddeeva V (1959) Computation methods of linear algebra. Dover translation from the Russian 1950 edition, New York

Fenner T, Loizou G (1974) Some new bounds on the condition numbers of optimally scaled matrices. J Assoc Comp Mach 21: 514–524

Ferguson A, Dantzig G (1956) The allocation of aircraft to routes-An example of linear programming under uncertain demand. Management Sc. 3: 45–73

Fiacco A (1983) Mathematical programming with data perturbations. North-Holland, New York

Fitzgerald K (1970) Error estimates for the solution of linear algebraic systems. J Res Nat Bur Standards Sec B 74B: 251–310

Fletcher R (1969) A review of methods for unconstrained optimization, Optimization Symp., Univ. of Keele, England. Academic Press, London

Fletcher R (1975) On the iterative refinement of least-squares solutions. J Amer Statist Assoc 70: 109–112

Forsythe G, Moler C (1967) Computer solutions of linear algebraic systems. Prentice-Hall, N.J.

Forsythe G, Malcolm M, Moler C (1977) Computer methods for mathematical computations. Prentice-Hall, N.J.

Franklin J (1968) Matrix theory. Prentice-Hall, N.J.

Fröberg C (1969) Introduction to numerical analysis, 2nd ed. Addison-Wesley, Reading, Mass.

Gal T, Nedoma J (1972) Multiparametric linear programming. Management Sc 18: 406–422

Galton F (1886) Family likeness in Stature. Proc. Roy Soc London 40: 42–72

Garloff J (1980) Totally nonnegative interval matrices. In: Nickel K (ed) Interval mathematics. Academic Press New York

Garloff J (1985) Interval mathematics, a bibliography, Freiburger Intervall-Berichte 85/6, Institut für Angewandte Mathematik, Universität Freiburg

Gass S (1969) Linear programming, methods and applications, 3rd ed., McGraw-Hill, Tokyo

Gass S, Saaty T (1955): Parametric objective function(Part 2)-generalization. Opns Res 3: 395–401

Gay D (1982) Solving interval linear equations. Siam J Numer Anal 19: 858–870

Geary R (1930) The frequency distribution of the quotient of two normal variates. J Roy Statist Soc 93: 442–446

Geurts A (1982) A contribution to the theory of condition. Numer Math. 39: 85–96

Golub G (1965) Numerical methods for solving linear least-squares problem. Numer Math 7: 206–216

Golub G, Kahan W (1965) Calculating the singular values and pseudo-inverse of a matrix. Siam J Numer Anal 2: 205–224

Golub G, Wilkinson J (1966) Note on the iterative refinement of least-squares solution. Numer Math 9: 139–148

Golub G, Reinsch C (1970) Singular value decomposition and least-squares solutions. Numer Math 14: 403–420

Golub G, Nash S, Van Loan C (1979) A Hessenberg-Schur method for the problem $AX + XB = C$, IEEE Trans Autom Contr AC-24, 909–913

Golub G, Van Loan C (1980) An analysis of the total least-squares problem. Siam J Numer Anal 17: 883–893

Golub G, Plemmons R (1980) Large-scale geodetic least-squares adjustment by dissection and orthogonal decomposition. Linear Alg and its Appl 34: 3–27

Golub G, Van Loan C (1983) Matrix computations. The John-Hopkins Univ. Press, Baltimore, Maryland

Graybill F, Bowden D (1967) Linear segment confidence bands for simple linear models. J Amer Statist Assoc 62: 403–408

Grimes R, Lewis J (1981) Condition number estimation for sparse matrices. Siam J Sci Stat Comput 2: 384–388

Hager W (1984) Condition estimates. Siam J Sci Stat Comput 5: 311–316

Halperin M (1961) Fitting of straight lines and prediction when both variables are subject to error. J Amer Statist Assoc 56: 657–669

Hammersley J, Handscomb D (1964) Monte Carlo methods. Methuen, London

Hamming R (1971) Introduction to applied numerical analysis. McGraw-Hill, New lork

Hansen E (1965) Interval arithmetic in matrix computations, Part I. Siam J Numer Anal 2: 308–320

Hansen E, Smith R (1967) Interval arithmetic in matrix computations, Part II. Siam J Numer Anal 4: 1–9

Hansen E (1969) On linear algebraic equations with interval coefficients. In: Hansen E (ed) Topics in interval analysis. Clarendon Press, Oxford

Hanson M (1960) Errors and stochastic variations in linear programming. Aust J Statist 2: 41–46

Hartfiel D (1980) Concerning the solution set of $Ax = b$ where $P \leqq A \leqq Q$ and $p \leqq b \leqq q$. Numer Math 35: 355–359

Heath M (1984) Numerical methods for large sparse linear least-squares problems. Siam J Sci Stat Comput 5: 497–513

Hocking R (1983) Developments in linear regression methodology: 1959–1982. Technometrics 25: 219–230

Hodges S, Moore P (1972) Data uncertainties and least squares regression. Appl Statist 21: 185–195

Hoffman A (1952) On approximate solutions of systems of linear inequalities. J Res Nat Bur Stand 49: 263–265

Hotelling H (1943) Some new methods in matrix calculations. Annals of Math Statist 14: 1–33

Householder A (1958) The approximate solution of matrix problems. J Assoc Comp Mach 5: 205–243

Householder A (1964) The theory of matrices in numerical analysis. Blaisdell, Col.

Jahn K (1974) Eine Theorie der Gleichungssysteme mit Intervall-Koeffizienten. ZAMM 54: 405–412

Johnston J (1972) Econometric methods, 2nd ed. McGraw-Hill, Tokyo

Jonckheere E (1984) New bound on the sensitivity of the solution of the Lyapunov equation. Linear Alg. and its Appl. 60: 57–64

Kahan W (1958) Gauss-Siedel methods for solving large systems of linear equations, Doctoral Thesis. Univ. of Toronto

Kahan W (1966) Numerical linear Algebra. Canad Math Bull 9: 757–801

Kallberg J, White R, Ziemba W (1982) Short-term financial planning under uncertainty. Management Sc 28: 670–682

Kendall M, Stuart A (1961) The advanced theory of statistics, vol. 2: Inference and relationship. Charles Griffin, London

Ketellapper R (1983) On estimating parameters in a simple linear errors-in-variables model. Technometrics 25: 43–47

Kovarik Z (1977) Minimal compatible solutions of linear equations. Linear Alg. and its Appl. 17: 95–106

Kramarz L (1981) Algebraic perturbation methods for the solution of singular linear systems. Linear Alg. and its Appl. 36: 79–88

Kulisch U, Miranker W (1983) A new approach to scientific computation. Academic Press, New York

Kuperman I (1971) Approximate linear algebraic equations. Van Nostrand Reinhold Co., London

Langenhop C (1971) The Laurent expansion for a nearly singular matrix. Linear Alg. and its Appl. 4: 329–340

Langenhop C (1973) On the invertibility of a nearly singular matrix. Linear Alg. and its Appl. 7: 361–365

La Porte M, Vignes J (1975) Evaluation de l'incertitude sur la solution d'un système lineaire. Numer Math 24: 39–47

Larson J, Sameh A (1980) Algorithms for roundoff error analysis — A relative error approach. Computing 24: 275–297

Lata M, Mittal B (1976) A decomposition method for interval linear fractional programming. ZAMM 56: 153–159

Lawson C, Hanson R (1974) Solving least-squares problems. Prentice-Hall, N.J.

Lemeire F (1973) Bounds for condition numbers of triangular and trapezoid matrices. BIT 15: 58–64

Levenberg K (1944) A method for the solution of certain nonlinear problems in least-squares. Quart J Appl Math 2: 164–168

Levy H, Sarnat M (1977) Financial decision making under uncertainty. Academic Press, New York

Lindley D (1947) Regression lines and the linear functional relationship. J Roy Statist Soc, Ser B 9: 218–244

Ljung G, Box G (1980) Analysis of variance with autocorrelated observations. Scand J Statist 7: 172–180

Loizou G (1968) An empirical estimate of the relative error of the computed solution $\bar{x}$ of $Ax = b$. Comput J 11: 91–94

Lonseth A (1942) Systems of linear equations with coefficients subject to error. Ann Math Statist 13: 332–337

Lonseth A (1944) On relative errors in systems of linear equations. Ann Math Statist. 15: 323–325

Lonseth A (1947) The propagation of error in linear problems. Trans. Amer Math Soc 62: 193–313

Lötstedt P (1984) Solving the minimal least-squares problem subject to bounds on the variables. BIT 24: 206–224

Luecke G (1979) A numerical procedure for computing the Moore-Penrose inverse. Numer Math 32: 129–137

Mandasky A (1959) The fitting of straight lines when both variables are subject to error. J Amer Statist Assoc 54: 173–205

Mandasky A (1962) Methods of solution of linear programs under uncertainty. Opns Res 10: 463–471

Manteuffel T (1981) An interval analysis approach to rank determination in linear least-squares problems. Siam J Sci Stat Comput 2: 335–348

Marquardt D (1963) An algorithm for least-squares estimation of non-linear parameters. J Soc Ind Appl. Math 11: 431–441

Martin D (1975) On the continuity of the maximum in parametric linear programming. J Optim Theory and Appl. 17: 205–210

Maurin H (1965) Parametrization générale d'un programme lineaire. Rev France Rech Opér 8: 277–292

Mayer O (1968) Über die in der Intervallrechnung auftretenden Räume und einige Anwendungen. Dissertation, Universität Karlsruhe

Mayer O (1970) Algebraische und metrische Strukturen in der Intervallrechnung und einige Anwendungen. Computing 5: 144–162

McCarthy C, Strang G (1973) Optimal conditioning of matrices. Siam J Numer Anal 10: 370–388

McCormick S (1975) A uniform approach to gradient methods for linear operator equations. J Math Anal and Appl. 49: 275–285

McElroy F (1967) A necessary and sufficient condition that ordinary least-squares estimators be best linear unbiased. J Amer Statist Assoc 62: 1302–1304

Meinguet J (1969) On the estimation of significance. In: Hansen E (ed) Topics in interval analysis. Clarendon Press, Oxford

Miller W (1972) On an interval-arithmetic matrix method. BIT 12: 213–219

Miranda C (1941) Un'osservatione su un teorema di Brouwer. Boll Un Mat Ital Serie II: 3: 5–7

Moler C (1978) Three research problems in numerical linear algebra. In: Golub G, Oliger J (ed) Numerical analysis. Proc Symp Appl Math 22: 1–18

Moler C (1980) MATLAB user's guide, Tech Rep CS81-1. Dept. of Computer Sc. Univ. of New Mexico, Albuquerque

Mönch W (1978) Monotone Einschließung der Moore-Penrose Pseudoinversen einer Matrix. ZAMM 58: 67–74

Moore E (1920) On the reciprocal of the general algebraic matrix. Bull Amer Math Soc 26: 394–395

Moore R (1966) Interval analysis. Prentice-Hall, N.J.

Moore R (1969) Introduction to algebraic problems. In: Hansen E (ed) Topics in interval analysis. Clarendon Press, Oxford

Moore R (1979) Methods and applications of interval analysis. Siam studies in Appl. Math. Siam, Philadelphia
Moore R, Nashed M (1974) Approximations to generalized inverses of linear operators. Siam J Appl Math 27: 1–16
Moran P (1971) Estimating structural and functional relationships. J Multivariate Anal 1: 232–255
Murty K (1966) Two-stage linear program under uncertainty: A basic property of the optimal solution. Opns Res Center. Univ. of Calif. Berkeley
Nake F (1967) Über die Anzahl der reellen Lösungen zufälliger Gleichungssysteme. Dissertation. Technische Hochschule, Stuttgart
Nash J (1979) Compact numerical methods for computers: Linear algebra and function minimization. Adam Hilger, Bristol
Nashed M (1970) Steepest descent for singular linear operator equations. Siam J Numer Anal 7: 358–362
Nashed M (1976) Perturbations and approximations for generalized inverses and linear operator equations. In: Nashed M (ed) Generalized inverses and applications. Academic Press, New York
Nashed M, Engl H (1979) Random generalized inverses and approximate solutions of random operator equations. In: Bharucha-Reid A (ed) Approximate solution of random equations. North-Holland, New York
Nickel K (1975) Interval mathematics. Springer, Berlin, Heidelberg, New York (Lecture Notes in Computer Science, Vol. 29)
Nickel K (1977) Interval analysis. In: Jacobs D (ed) The state of the art in numerical analysis. Academic Press, London
Nickel K (1980) Interval mathematics. Academic Press, New York
Noble B (1969) Applied linear algebra. Prentice-Hall, N.J.
Noble B (1976) Methods for computing the Moore-Penrose generalized inverse, and related matters. In: Nashed M (ed) Generalized inverses and applications. Academic Press, New York
Nuding V, Wilhelm J (1972) Über Gleichungen und über Lösungen. ZAMM 52: T188–190
Oettli W, Prager W (1964) Compatibility of approximate solution of linear equations with given error bounds for coefficients and right-hand sides. Numer Math 6: 405–409
Oettli W (1965) On the solution set of a linear system with inaccurate coefficients. Siam J Numer Anal 2: 115–118
Oettli W, Prager W, Wilkinson J (1965): Admissible solutions of linear systems with not sharply defined coefficients. Siam J Numer Anal 2: 291–299
O'Leary D (1980) Estimating matrix condition numbers. Siam J Sci Stat Comput 1: 205–209
Oman S (1983) Regression estimation for a bounded response over a bounded region. Technometrics 25: 251–261
Ostrowski A (1950) Sur la variation de la matrice inverse d'une matrice donnée, C R Acad SC Paris 231: 1–3
Paige C (1973) An error analysis of a method for solving matrix equations. Math Comput 27: 355–359
Paige C (1979) Computer solution and perturbation analysis of generalized least-squares problems. Math Comput 33: 171–184
Papoulis A (1966) Perturbations of the natural frequencies and eigenvectors of a network. IEEE Trans. on circuit theory, CT-13: 188–195
Pease M (1965) Methods of matrix algebra. Academic Press, New York
Penrose R (1955) A generalized inverse for matrices. Proc Cambridge Philos Soc 51: 406–413
Peters G, Wilkinson J (1970) The least-squares problem and pseudo-inverses. Comput J 13: 309–316
Peters G, Wilkinson J (1979) Inverse iteration, ill-conditioned equations and Newton's method. Siam Rev 21: 339–360
Pierce D, Gray R (1982) Testing normality of errors in regression models. Biometrika 69: 233–236

Prekopa A (1966) On the probability distribution of the optimum of a random linear program. Siam J Control 4: 211–222

Raj B (1980) A monte Carlo study of small-sample properties of simultaneous equation estimators with normal and non-normal disturbances. J Amer Statist Assoc 75: 221–229

Rall L (1979) Perturbation methods for the solution of linear problems. In: Nashed M (ed) Functional analysis methods in numerical analysis. Springer, Berlin, Heidelberg, New York

Ralston A (1965): A first course in numerical analysis. McGraw-Hill, New York

Rao C (1965) Linear statistical inference and its applications. Wiley, New York

Rao C, Mitra S (1971) Generalized inverse of matrices and its applications, Wiley, New York

Rao C, Mitra S (1972) Generalized inverse of a matrix and its applications. Proc. Six*th* Berkeley Symp on Math Statist and Prob. Univ. of Calif. Press, Vol. I

Rao C (1972) Estimation of variance and covariance components in linear models. J Amer Statist Assoc 67: 112–115

Ramsey J (1969) Tests for specification errors in classical least-squares regression analysis. J Roy Statist Soc 31: 351–371

Rappaport A (1967) Sensitivity analysis in decision making. Accounting Rev XLII: 441–456

Ratschek H (1975) Nichtnumerische Aspekte der Intervall-arithmetic. In: Nickel K (ed) Interval mathematics. Springer, Berlin, Heidelberg, New York (Lecture Notes in Computer Science, Vol 29)

Ratschek H, Sauer W (1982) Linear interval equations. Computing 28: 105–115

Ribaric M, Vidav I (1969) Analytic properties of the inverse $A(z)^{-1}$ of an analytic linear operator valued function $A(z)$. Arch Rat Mech Anal 32: 298–310

Rice J (1966) A theory of condition. Siam J Numer Anal 3: 287–310

Rice J (1981) Matrix computations and mathematical software. McGraw-Hill, Tokyo

Rigal J, Gaches J (1967) On the compatibility of a given solution with the data of a linear system. J. Assoc Comp Mach 14: 543–548

Robers P, Ben-Israël A (1970) A suboptimization method for interval linear programming: A new method for linear programming. Linear Alg. and its Appl. 3: 383–405

Robertson H (1977) The accuracy of error estimates for systems of linear algebraic equations. J Inst Math Appl 20: 409–414

Robinson S (1973) Bounds for error in the solution set of a perturbed linear program. Linear Alg. and its Appl. 6: 69–81

Rohn J (1978) Correction of coefficients of the input-output model. ZAMM 58: T494–495

Rohn J (1981a) Interval linear systems with prescribed column sums. Linear Alg. and its Appl. 39: 143–148

Rohn J (1981b) Strong solvability of interval linear programming problems. Computing 26: 79–82

Rokne J (1978) Polynomial least-square interval approximation. Computing 20: 165–176

Rose N (1978) The Laurent expansion of a generalized resolvent with some applications. Siam J Math Anal 9: 751–757

Rosenthal R (1976) Sufficient conditions for insensitivity in linear models. Opns Res 24: 183–188

Rump S, Kaucher E (1980) Small bounds for the solution of systems of linear equations. Computing, Suppl 2: 157–164

Saaty T, Gass S (1954) Parametric objective function (Part I). Opns Res 2: 316–319

Saaty T (1959) Coefficient perturbation of a constrained extremum. J Opns Res Soc of Amer 7: 294–302

Salvadori M and Baron M (1966) Numerical methods in engineering. Prentice-Hall of India, New Delhi

Sandor P (1964) Some problems of ranging in linear programming. CORS J 2: 26–31

Seal H (1967) Studies in the history of probability and statistics XV. The historical development of the Gauss linear model. Biometrika 54: 1–24

Searle S (1971) Topics in variance components estimation. Biometrika 27: 1–76

Senechal L (1985) Taylor series expansions for functions of linear operators, 2*nd* Siam Conf. on Applied Linear Algebra. Raleigh, NC

Shetty C (1959a) Sensitivity analysis in linear programming. J Ind Eng X: 379–386
Shetty C (1959b) Solving linear programming problems with variable parameters. J Ind Eng X: 433–438
Simons E (1962) A note on parametric linear programming. Management Sc 8: 355–358
Skeel R (1979) Scaling for numerical stability in Gaussian elimination. J Assoc Comput Mach 26: 494–526
Skelboe S (1979) True worst-case analysis of linear electrical circuits by interval arithmetic. IEEE Trans. on circuits and systems, CAS-26: 874–879
Smith B, Boyle J, Ikebe Y, Klema V, Moler C (1970) Matrix Eigensystem Routines EISPACK Guide. Springer, Berlin, Heidelberg, New York (Lecture Notes in Computer Science, Vol 6)
Spellucci P, Krier N (1976) Ein Verfahren zur Behandlung von Ausgleichsaufgaben mit Intervallkoeffizienten. Computing 17: 207–218
Sprent P (1969) Models in regression and related topics. Methuen, London
Stenlund H, Westlund A (1975): A monte Carlo study of simple random sampling from a finite population. Scand J Statist 2: 106–108
Steuer R (1976) Multiple objective linear programming with interval criterion weights. Management Sc 23: 305–316
Steuer R (1981) Algorithms for linear programming problems with interval objective function coefficients. Math of Opns Res 6: 333–348
Stewart G (1969): On the continuity of the generalized inverse. Siam J Appl Math 17: 33–45
Stewart G (1971) Error bounds for approximate invariant subspaces of closed linear operators. Siam J Numer Anal 8: 796–808
Stewart G (1973a) Error and perturbation bounds for subspaces associated with certain eigenvalue problems. Siam Rev 15: 727–764
Stewart G (1973b) Introduction to matrix computations. Academic Press, New York
Stewart G (1977) Perturbation bounds for the *QR* factorization of a matrix. Siam J Numer Anal 14: 509–518
Stewart G (1977) On the perturbation of pseudo-inverses, projections and linear least-squares problems. Siam Rev 19: 634–661
Stewart G (1979) A note on the perturbation of singular values. Linear Alg. and its Appl 28: 213–216
Stewart G (1980) The efficient generation of random orthogonal matrices with an application to condition estimators. Siam J Numer Anal 17: 403–409
Stewart N (1973) Interval arithmetic for guaranteed bounds in linear programming. J Optim Theory and Appl 12: 1–5
Stigler S (1981) Gauss and the invention of least squares. Annals of statistics 9: 465–474
Stoer J (1964) On the characterization of least upper bound norms in matrix space. Numer Math 6: 302–314
Stoer J, Bulirsch R (1980) Introduction to numerical analysis. Springer, Berlin, Heidelberg, New York
Sun J (1983) Perturbation analysis for the generalized singular value problem. Siam J Numer Anal 20: 611–625
Temes G, LaPatra J (1977) Network analysis and synthesis. McGraw-Hill, New York
Thieler P (1975) Verbesserung von Fehlerschranken bei iterativer Matrizeninversion. In: Nickel K (ed) Interval mathematics. Springer, Berlin, Heidelberg, New York (Lecture Notes in Computer, Vol. 29)
Thomas H, Revelle R (1966) On the efficient use of High Aswan Dam for hydropower and irrigation. Management Sc 12: B297–311
Thompson R (1976) The behavior of eigenvalues and singular values under perturbations of restricted rank. Linear Alg. and its Appl. 13: 69–78
Turing A (1948) Rounding-off errors in matrix processes. Quart J Mech Appl Math 1: 287–308
Usmani R, Chebib F (1978) A note on the least-square linear approximation. BIT 18: 112–115

van de Panne C, Bosje P (1962) Sensitivity analysis of cost coefficient estimates: The case of linear decision rules for employment and production. Management Sc 9: 82–107
van der Sluis A (1969) Condition numbers and equilibrium of matrices. Numer Math 14: 14–23
van der Sluis A (1970a) Stability of solutions of linear algebraic systems. Numer Math 14: 246–251
van der Sluis A (1970b) Condition, equilibration and pivoting in linear algebraic systems. Numer Math 15: 74–86
van der Sluis A (1975) Stability of the solutions of linear least squares problems. Numer Math 23: 241–254
Varah J (1973) On the numerical solution of ill-conditioned linear systems with applications to ill-posed problems. Siam J Numer Anal 10: 257–267
Varah J (1975) A lower bound for the smallest singular value of a matrix. Linear Alg. and its Appl. 11: 3–5
Varah J (1979) On the separation of two matrices. Siam J Numer Anal 16: 216–222
Varga R (1962) Matrix iterative analysis, Prentice-Hall, N.J.
Varga R (1976) On diagonal dominance arguments for bounding $\|A^{-1}\|$, Linear Alg. and its Appl. 14: 211–217
von Neumann (1937) Some matrix inequalities and metrization of matrix-space. Inst Mat Mech Tomsk Gos Univ. 1: 286–299
von Neumann J, Goldstine H (1947) Numerical inverting of matrices of high order. Bull Amer Math Soc 53: 1021–1099
Wald A (1940) The fitting of straight lines if both variables are subject to error. Annals of Mathematical Statistics XI: 284–300
Webb K (1962) Some aspects of the Saaty linear programming sensitivity equation. Opns Res 10: 266–267
Wedderburn J (1934) Lectures on matrices. Amer Math Soc. New York
Wedin P (1972) Perturbation bounds in connection with singular value decomposition. BIT 12: 99–111
Wedin P (1973) Perturbation theory for pseudo-inverses. BIT 13: 217–232
Wets R (1964) Programming under uncertainty, the complete problem. Boeing Scientific Res Lab D1-82-0739
Wets R (1965a) Programming under uncertainty — The equivalent convex problem. Boeing Scientific Res Lab D1-82-0411
Wets R (1965b) Programming under uncertainty — The solution set. Boeing Scientific Res Lab D1-82-0464
Wilkinson J (1960) Error analysis of floating-point computation. Numer Math 2: 319–340
Wilkinson J (1961) Error analysis of direct methods of matrix inversion. J Assoc Comp Mach 8: 281–330
Wilkinson J (1963) Rounding errors in algebraic processes. Prentice-Hall, N.J.
Wilkinson J (1965) The algebraic eigenvalue problem. Clarendon Press, Oxford
Wilkinson J (1971) Modern error analysis. Siam Rev. 13: 548–568
Wilkinson J, Reinsch C (1971) Handbook for automatic computation, Vol. II: Linear algebra. Springer, Berlin, Heidelberg, New York (Grundlehren d. mathematischen Wissenschaft Bd. 86)
Williams A (1963) Marginal values in linear programming. Siam J Appl Math 11: 82–94
Wolfe P (1965) Error in the solution of linear programming problems. In: Rall L (ed) Errors in digital computation, Vol. 2, Wiley, New York
Wongwises P (1975) Experimentelle Untersuchungen zur Numerischen Auflösung von Linearen Gleichungssystemen mit Fehlererfassung. In: Nickel K (ed) Interval mathematics. Springer Berlin, Heidelberg, New York (Lecture Notes in Computer Science, Vol 29)
Wozniakowski H (1978) Round-off error analysis of iterations for large linear systems. Numer Math 30: 301–314
Yamamoto T (1975) On the distribution of errors in the iterative solution of a system of linear equations. Numer Math 24: 71–79
Yamamoto T (1976) On the distribution of errors in the iterative solution of a system of linear equations. II Numer Math 25: 461–463

Yamamoto T (1981) Componentwise error estimates for approximate solutions of systems of equations. Lecture notes in Numer Appl Math Anal 3: 1–22

Yamamoto T (1984) Error bounds for approximate solutions of systems of equations. Japan J of Appl Math 1: 157–171

Yau S (1964) The sensitivity of a traffic Network. J Franklin Inst. 278: 371–382

Yedavalli R (1985) Stability analysis of interval matrices-Another sufficient condition, 2nd Siam Conf. on Appl Linear Algebra, Raleigh, NC

Young D (1950) Iterative methods for solving partial difference equations of elliptic type. Doctoral Thesis, Harvard Univ.

Young D (1954) Iterative methods for solving partial difference equations of elliptic type. Trans Amer Math Soc 76: 92–111

Young D (1971) Iterative solution of large linear systems. Academic Press, London

Young D, Gregory R (1973) A survey of numerical mathematics, Vol. 2. Addison-Wesley, Reading Mass.

Yu P, Zeleny M (1976) Linear multiparametric programming by multicriteria simplex method. Management Sc 23: 159–170

Zereick L, Amer R, Deif A (1983) Network tuning via mixed integer programming. In: Mitra, S. (ed) Proc. IEEE Int. Symp. on circuits and systems (ISCAS)

Zlamal J (1977) A measure for ill-conditioning of matrices in interval arithmetic. Computing 19: 149–155

Index